Studies in Computational Intelligence

Volume 1222

The series "Studies in Computational Intelligence" (SCI) publishes new developments and advances in the various areas of computational intelligence—quickly and with a high quality. The intent is to cover the theory, applications, and design methods of computational intelligence, as embedded in the fields of engineering, computer science, physics and life sciences, as well as the methodologies behind them. The series contains monographs, lecture notes and edited volumes in computational intelligence spanning the areas of neural networks, connectionist systems, genetic algorithms, evolutionary computation, artificial intelligence, cellular automata, self-organizing systems, soft computing, fuzzy systems, and hybrid intelligent systems. Of particular value to both the contributors and the readership are the short publication timeframe and the world-wide distribution, which enable both wide and rapid dissemination of research output.

Indexed by SCOPUS, DBLP, WTI AG (Switzerland), zbMATH, SCImago.

All books published in the series are submitted for consideration in Web of Science.

László T. Kóczy · Jesús Medina ·
Piotr A. Kowalski · Eloísa Ramírez-Poussa
Editors

Computational Intelligence and Mathematics for Tackling Complex Problems 6

Springer

Editors
László T. Kóczy
Faculty of Engineering Sciences
Széchenyi István University
Győr, Hungary

Budapest University of Technology
and Economics
Budapest, Hungary

Piotr A. Kowalski
Faculty of Physics and Applied Computer
Science
AGH University of Kraków
Kraków, Poland

Systems Research Institute
Polish Academy of Sciences
Warsaw, Poland

Jesús Medina
Department of Mathematics, Faculty
of Science
University of Cádiz
Cádiz, Spain

Eloísa Ramírez-Poussa
Faculty of Engineering, Department
of Mathematics
University of Cádiz
Cádiz, Spain

ISSN 1860-949X ISSN 1860-9503 (electronic)
Studies in Computational Intelligence
ISBN 978-3-031-97878-4 ISBN 978-3-031-97879-1 (eBook)
https://doi.org/10.1007/978-3-031-97879-1

This Springer imprint is published by the registered company Springer Nature Switzerland AG
The registered company address is: Gewerbestrasse 11, 6330 Cham, Switzerland

Preface

Artificial intelligence appears nowadays very much in the media, and also in the practice. In addition, every day, thousands of automatic intelligent systems for decision-making are designed and applied around the world. In their design many advanced techniques based on mathematics and computational intelligence are considered and used. Therefore, it is fundamental to continuously work on novel approaches in these areas and research models and algorithms with foundation in the mutual relationship of these two fields.

This volume of the series is focused on collecting some of the best results with this goal, dealing with the important and efficient combination of mathematics and computational intelligence, with a stress of fuzzy systems, and fuzzy logic. Naturally, an important part of this volume focuses on the applications of these both areas, taking a special interest in the development and application of mathematics, artificial intelligence and automatic reasoning tools applicable to Digital Forensics, also following the main objective of the COST Action DigForASP-CA17124 (digforasp.uca.es).

The collected works are extended and revised versions of selected papers presented at the conference 15th European Symposium on Computational Intelligence and Mathematics held on 12–15 May 2024, in Krakow (Poland). Moreover, the technical program of this conference included four excellent keynote speeches, given by Prof. Janusz Kacprzyk (Professor of Computer Science at the Systems Research Institute, Polish Academy of Sciences, WIT-Warsaw School of Information Technology, AGH University of Science and Technology in Cracow, and Professor of Automatic Control at PIAP-Industrial Institute of Automation and Measurements in Warsaw), Prof. Radu-Emil Precup (Professor in the Department of Automation and Applied Informatics of the Politehnica University of Timisoara in Romania), Prof. Viviana Mascardi (Associate Professor in Computer Science at the University of Genova, Italy, DIBRIS), and Prof. Piotr Kowalski (Professor at the AGH University of Science and Technology, working at the Faculty of Physics and Applied Computer Science, as well as at the Systems Research Institute of the Polish Academy of Sciences).

We acknowledge the work given by the Program Committee members, reviewers, authors, local organizers and the support of the project PID2019108991GB-I00 funded by MICIU/AEI/10.13039/501100011033, the project PID2022137620NB-I00 funded by MICIU/AEI/10.13039/501100011033 and FEDER, UE, the grant TED2021-129748B-I00 funded by MCIN/AEI/10.13039/501100011033 and European Union NextGenerationEU/PRTR, and the National Research, Development and Innovation Office (Hungary) grant no. K124055. In what follows, the chapters will be briefly described.

The first chapter "A Note on Aggregation of T-Fuzzy Subgroups" provides an answer to the question of the preservation of fuzzy subgroup structure for some special classes of binary aggregation operators. The authors develop a new method for n-ary aggregation applicable to fuzzy subgroups regardless of the cardinality, based on n-ary aggregation operators generated by a family of binary aggregation operators.

The second chapter "Independent Subcontexts in Fuzzy Formal Concept Analysis" addresses the factorization problem within the theory of Formal Concept Analysis. In this paper, the authors formalize the notion of independent subcontext of a L-fuzzy context and introduce some properties that establish the existence of independent subcontexts in the residuated lattice framework.

The third chapter "Two Sufficient Conditions for Having a Maximal Solution to Fuzzy Relational Equations" studies the solvability of fuzzy relational equations. In particular, the authors prove the existence of a maximal solution to some fuzzy relational equations, when the codomain lattice is complete and meet-continuous.

The fourth chapter "Towards Compatible Subcontext by Means of Classical Attribute Reduction" is framed in the theory of formal concept analysis. It explores sufficient conditions to obtain compatible contexts of a formal contexts. Compatible contexts are useful to reduce the computational complexity in the analysis of data, which also can complement other reduction mechanisms considered in this mathematical theory.

The fifth chapter "Extension of The Time Dependent Travelling Salesman Problem Model with Interval-Valued Fuzzy Soft Sets with Arithmetic Mean Method" studies the Traveling Salesman Problem, which is a known NP-hard graph search problem. The authors introduce a more realistic novel fuzzy-based extension, the Interval-Valued Fuzzy Soft Set for the Time Dependent Traveling Sales-man Problem, which is a useful extension and a practical alternative model of the original abstract problem.

The sixth chapter "On Aggregation Operators and Utility Functions with Truncation Property" presents an overview of results that consider the conditional distributivity of aggregation operators, namely T-uninorms and nullnorms, over t-conorms and an analysis of the role of obtained distributive pairs in constructions of utility functions with a truncation property.

The seventh chapter "Categories of L-Fuzzy Morphological Operators on L-Fuzzy Groups and LR-Fuzzy Homomorphisms" is frames in mathematical morphology. In particular, in this work, the authors introduce two categories based on completely distributive lattice with the goal of developing a categorical approach to classical fuzzy mathematical morphology.

The eighth chapter "Message Passing Graph Neural Network for Seeds Classification" explores machine learning techniques within the field of High Energy Physics. The authors combine the traditional Combinatorial Kalman Filter with a GNN-based solution to improve the analysis of particle collisions at the Large Hadron Collider.

The ninth chapter "An Interval-Valued Fuzzy Soft Sets Based Decision Support Model for Route Optimization" addresses another known NP-hard graph search problem, route optimization. In this paper, the authors propose a novel route optimization under road uncertainties using Interval-Valued Fuzzy Soft Sets. The considered approach is a practical and closer to reality estimation for non-deterministic factors of the original abstract route optimization problem.

The tenth chapter "Semi-automatic Knowledge Representation and Reasoning on Vague Crime Concepts" tackles the problem of the formal representation of vagueness in the legal domain. With this goal, the authors consider Answer Set Programming as modeling language. This work starts the first steps towards the attainment of a legal reasoning system capable of evolving by doing fully automatic learning from sentences.

The eleventh chapter "Empowering Emotional Behavior Trees with Neural Computation for Digital Forensic" makes use of behavior trees to model the empathetic behavior of virtual agents interacting with humans. This allows the specification of various kinds of empathy for the agents to be deployed for user support in critical fields such as digital forensics.

The twelfth chapter "Legal and Technical Challenges of AI in the Field of Criminal Investigations" addresses the need for ethical considerations and legal frameworks to guide the responsible development and deployment of artificial intelligence tools in digital forensics. Three key issues in criminal investigations are explored: the discriminatory biases originated by databases and algorithms, the lack of traceability in artificial intelligence based decisions and the creation of anonymisation systems to safeguard individuals' privacy rights.

The thirteenth chapter "Fingerprint Revolution: Unleashing the Potential of Modified Bacterial Memetic Evolution for a Paradigm Shift in Fingerprint Recognition and Optimization" proposes a new approach to improve fingerprint matching outcomes by using modified optimization algorithms. This approach is based on innovative methodologies, aiming to remove undesired elements of fingerprint images during the matching process and reduce the algorithm complexity time, by segmenting fingerprint images into three pieces.

The fourteenth chapter "Comparison of Text Similarity Techniques for Power of Attorney Clauses for Polish Banks" carries out a comparative study about text similarity techniques given in the literature. The effectuality of the text similarity techniques has been analyzed by their application to power of attorney clauses for Polish banks.

The fifteenth chapter "Belief Change: Axiomatic Characterization of KM-Erasure" presents a characterization of erasure for belief sets from both an axiomatic and semantic perspective. In addition, the relationship between erasure and update through Levi and Harper identities is studied.

The sixteenth chapter "Stable Reasoning, ASP and the Interrogative Model of Inquiry" obtains a close agreement between the interrogative model and question answering in answer set programming, through the extension of the fixed point property of propositional equilibrium models to the first order case.

The seventeenth chapter "Hypergraph Logic Program Representation Versus Stratified Programs" provides a hypergraph representation of a multiadjoint normal logic program which characterizes the given program. An optimal stratification of the considered program is defined from its condensation graph. This fact allows to obtain the least model of the subprogram associated with each stratum in parallel.

The eighteenth chapter "Enhancement of Discrete Bacterial Memetic Evolutionary Algorithm for Solving the Travelling Repairman Problem" combines, in particular, the Discrete Bacterial Memetic Evolutionary Algorithm with an original method for generating the initial individual candidate solution, by means of Genetic Algorithm.

The nineteenth chapter "Simulated Annealing and Bacterial Foraging for Probabilistic Neural Network Parameters Adjustment" shows a novel performance comparison of Probabilistic Neural Networks training optimization between the Plug-in technique and two metaheuristics approaches such as Simulated Annealing and Bacterial Foraging Optimization.

The twentieth chapter "Inheritance of Completeness Between Systems of Strong and Weak Implications" compares weak and strong attribute implications obtaining interesting theoretical properties. Specifically, the inheritance of completeness between systems of strong and weak implications is studied.

The twenty-first chapter "Migrative Properties for Triangular Conorms and Fuzzy Implications" studies the migrativity and cross-migrativity of tconorms over fuzzy implications, including new motivations, features and analyzing the relationship among both properties.

In the last chapter "Masonry Strength Assessment Based on Fuzzy Signature Model", fuzzy signatures are used to determine the compressive strength of masonry based on groups of related measured values. Fuzzy arithmetic multiplication based on Zadeh's extension principle, the scalar product and the n-th root of fuzzy numbers are applied to perform the modeling and the computations.

Győr/Budapest, Hungary — László T. Kóczy
Cádiz, Spain — Jesús Medina
Kraków/Warsaw, Poland — Piotr A. Kowalski
Cádiz, Spain — Eloísa Ramírez-Poussa

Keynote Speaker

Sensitivity Analysis as a Method for Explaining AI (XAI) in the Artificial Neural Networks Domain

Abstract The field of Explainable Artificial Intelligence (XAI) has gained prominence as the adoption of complex models, such as Artificial Neural Networks (ANNs), continues to grow. Sensitivity Analysis emerges as a pivotal method within the XAI framework, offering a systematic approach to unravelling the intricate inner workings of ANNs. This invited talk delves into the significance of Sensitivity Analysis as a method for explaining AI in the domain of Artificial Neural Networks. The presentation explores how sensitivity analysis techniques contribute to transparency, interpretability, and trust in AI systems, shedding light on the factors that influence model predictions. Through illustrative examples and case studies, the talk aims to provide valuable insights into the practical application of sensitivity analysis, bridging the gap between the complex nature of ANNs and the need for comprehensible AI systems.

The talk will showcase both global and local approaches of Sensitivity Analysis, providing a comprehensive understanding of how these methods can be employed to dissect the decision-making processes within ANNs. Attendees will gain insights into the application of Sensitivity Analysis to various types of neural networks, including Multilayer Perceptrons (MLPs), Probabilistic Neural Networks (PNNs), and Convolutional Neural Networks (CNNs). Through illustrative examples and case studies, the presentation aims to demonstrate the versatility of sensitivity analysis in dissecting the complex structures of different neural network architectures.

By focusing on specific neural network models, namely MLPs, PNNs, and CNNs, the talk will highlight the adaptability of sensitivity analysis across diverse AI applications. The presentation underlines a deeper understanding of how global and local sensitivity analysis techniques can enhance the interpretability and explainability of ANNs, contributing to the responsible deployment of AI systems.

This work was partially supported by the program "Excellence initiative—research university" for the AGH University of Krakow and by Grant for Statutory Activity from the Faculty of Physics and Applied Computer Science of the AGH University of Krakow.

Piotr A. Kowalski
Faculty of Physics and Applied Computer Science
AGH University of Kraków
Kraków, Poland

Systems Research Institute
Polish Academy of Sciences
Warsaw, Poland

An Approach to Consensus and Dissensus-Driven Decision Making Under Fuzziness

Abstract We advocate a new way of decision making under fuzzy preferences and fuzzy majorities the essence of which is that the consensus-driven paradigm, which is very often employed, is not followed as it can often prohibit reaching innovative decisions.

First, we briefly review main developments in the broadly perceived group decision making and indicate the main general approaches, namely via: 1) Unanimous decisions when all agents agree without reservations; 2) consensus in which each agent agrees to give his/her consent to the decision reached, even if it would not be perfect, but acceptable, and is open to modify his or her testimonies; 3) majority rule when, e.g., more than a half, at least 2/3, etc., of the group votes in favor; 4) expert in which a special agent, an expert, is chosen for running the decision making process; 5) executive in which a special high level agent makes the decision with little or none involvement of other group members; 6) default in which a decision is made as needed, without any analysis.

We analyze some new directions in social sciences, cognitive science, psychology, decision theory, etc. in which there is an explicit critique of consensus as a viable and effective and efficient way of making group decisions. Basically, the argument is that "consensus is the quickest way to kill innovation." This argument is raised by many authors, and there are some attempts to use some more formal analyses.

We also use some results of the so-called entrepreneurial action theory which also mentions that the most interesting forms of entrepreneurship involve ideas that contradict prevailing wisdom, and opinions. These behaviors may be irrational and seem impossible but may lead to innovative, even revolutionary outcomes.

We propose a new model of dissensus-driven group decision making under fuzzy preferences and fuzzy majorities. We redefine the solution concepts along the line of the fuzzy cores and fuzzy consensus winners. We mention examples

Janusz Kacprzyk
Systems Research Institute
Polish Academy of Sciences
Warsaw, Poland

Applications of Metaheuristic Algorithms to Fuzzy Control and Model Building, Learning-Based Control, and Mobile Robot Navigation

Abstract An optimization problem finds the best (i.e., optimal) solution among all feasible solutions. An optimization problem consists of two key components: the objective function and the constraints, which are optional. The objective function evaluates and compares solutions in the context of all feasible solutions by computing the desired quantity to be minimized or maximized. Constraints can be added to limit the possible values for the variables of the objective function and possibly to link these variables.

The optimization algorithms find the solutions to the optimization problems (i.e., the optimal solutions) by trying variations of the initial solution and using the information gained to improve the solution. This solution finding can also be considered as learning, which is a popular topic nowadays. The complexity of classical algorithms is very high, which requires rather large amount of computation. Therefore, alternative algorithms with lower complexity are appreciated. Metaheuristic algorithms for finding optimal solutions have become very popular because they are much better in terms of efficiency and complexity than classical algorithms.

This presentation highlights some of the results obtained by the Process Control Group of the Politehnica University of Timisoara, Romania. The presentation will focus on representative applications implemented in our labs, with real-time validation against experimental results. The results highlighted here include various laboratory equipment such as pendulum crane systems, multi-tank systems, servo systems, twin rotor aerodynamic systems, magnetic levitation systems, anti-lock braking systems, mobile robots, magnetic levitation systems, active mass damper systems, and shape memory alloy systems.

The scope of development of these metaheuristic algorithms is to solve optimization problems involving tuning of low-cost fuzzy controllers, tuning of fuzzy models, reinforcement-based control in various schemes including adaptive ones, and solving optimization problems specific to mobile robot navigation.

This work was supported by a grant of the Ministry of Research, Innovation and Digitization, CNCS/CCCDI–UEFISCDI, project number ERANET-ENUAC-e-MATS, within PNCDI IV.

Radu-Emil Precup
Politehnica University of Timisoara (UPT)
Timisoara, Romania

Logic Programming and Legal Reasoning: the Past and the Future

Abstract Logic programming is an extremely powerful tool for legal reasoning due to its declarative, rule-based nature, suitable for modeling in a natural and understandable way legal rules and constraints. In this talk, I will discuss the advantages and disadvantages of exploiting logic programming in the legal domain, and I will provide examples of use in different logic programming languages.

Viviana Mascardi
University of Genova
Genoa, Italy

Contents

A Note on Aggregation of T-Fuzzy Subgroups

Ivana Štajner-Papuga and Andreja Tepavčević

Abstract This paper aims to provide an answer to the question of the preservation of fuzzy subgroup structure for some special classes of aggregation operators. Both binary and n-ary aggregation of fuzzy subgroups are considered. The main contribution of the approach proposed in this paper is the development of a new method for n-ary aggregation applicable to fuzzy subgroups regardless of the cardinality, based on n-ary aggregation operators generated by a family of binary aggregation operators.

Keywords Fuzzy subgroups · Aggregation operator

1 Introduction

One of the contemporary lines of research in the field of mathematics, both theoretical and applied, is directed toward modeling problems of decision-making. Fusion of data is one of the essential aspects of the majority of recently proposed models, and aggregation operators play an important role in this process. Therefore, aggregation operators and their properties are notions of interest. Since modeling problems from real-life backgrounds often requires dealing with imprecise data, it is necessary to merge fuzzy logic with the aggregation process. Having all this in mind, it is

This research was supported by the Science Fund of the Republic of Serbia, # Grant No. 6565, Advanced Techniques of Mathematical Aggregation and Approximative Equations Solving in Digital Operational Research-AT-MATADOR.

I. Štajner-Papuga (✉) · A. Tepavčević
Department of Mathematics and Informatics, Faculty of Sciences, University of Novi Sad, Novi Sad, Serbia
e-mail: ivana.stajner-papuga@dmi.uns.ac.rs

A. Tepavčević
e-mail: andreja@dmi.uns.ac.rs

A. Tepavčević
Mathematical Institute SANU, Belgrade, Serbia

L. T. Kóczy et al. (eds.), *Computational Intelligence and Mathematics for Tackling Complex Problems 6*, Studies in Computational Intelligence 1222,
https://doi.org/10.1007/978-3-031-97879-1_1

quite clear why the preservation of fuzzy structures during an aggregation process is an issue worth investigating. Recently, several results have been published about aggregations of fuzzy subgroups. The main question is when the aggregation of fuzzy subgroups is again a fuzzy subgroup. Such problems have been tackled in papers [1, 2, 8]. In [2], the question of when the aggregation of two fuzzy subgroups provides a fuzzy subgroup is solved for fuzzy subgroups with the cardinality of a prime number, while in [1, 8] similar questions are solved for aggregation of n fuzzy subgroups. In paper [8], preservation of the structure of T-subgroup of groups under aggregation operator is investigated in the case of groups whose subgroup lattice is a chain.

This short note provides an answer to the question of the preservation of fuzzy subgroup structure for some special classes of binary aggregation operators. Also, this result is extended to n-ary aggregation operators by the method from [4]. In [4], a special type of n-ary lattice-valued aggregation operator that is associated with a family of binary lattice-valued aggregation operators was introduced. Hence, this paper addresses the problem of aggregation of groups only for special binary aggregation operators from [4], and the results for n-ary aggregation operators are a consequence.

2 Preliminary Notions

This section consists of an overview of some well-known notions that are essential for understanding the main topic of this paper, such as t-norm, T-fuzzy subgroup, and aggregation operators [3, 5–7]. Also, some results from [4] that are the base for the presented research are given.

- A **t-norm** $T : [0, 1]^2 \rightarrow [0, 1]$ is a commutative, associative, and monotone operation on $[0, 1]^2$, such that 1 is its neutral element. The largest t-norm is the minimum, i.e., $T_M(x, y) = \min\{x, y\}$, and the smallest one is the so-called drastic t-norm of the following form

$$T_D(x, y) = \begin{cases} 0 & \text{for } \max\{x, y\} < 1 \\ \min\{x, y\} & \text{for } \max\{x, y\} = 1. \end{cases}$$

- Let $(G, *, ^{-1}, e)$ be a group, where $*$ is the binary operation, $^{-1}$ the unary operation and e the neutral element. A mapping $\mu : G \rightarrow [0, 1]$ is a **T-fuzzy subgroup** of a group G for a t-norm T, if

$$\mu(e) = 1, \quad \mu(x^{-1}) = \mu(x) \quad \text{and} \quad T(\mu(x), \mu(y)) \leq \mu(x * y).$$

- An **n-ary aggregation operator** $A : [0, 1]^n \rightarrow [0, 1]$ for some $n \in \mathbb{N}$, where $\mathbb{N}$ is the set of natural numbers, is an n-ary operation on $[0, 1]$, such that $A(0, 0, \ldots, 0) = 0$, $A(1, \ldots, 1) = 1$, and from $x_1 \leq y_1, x_2 \leq y_2, \ldots, x_n \leq y_n$ it follows that $A(x_1, x_2, \ldots, x_n) \leq A(y_1, y_2, \ldots, y_n)$. An aggregation operation is disjunctive if $\max\{x_1, x_2, \ldots, x_n\} \leq A(x_1, x_2, \ldots, x_n)$.

- A **binary aggregation operator** is obtained for $n = 2$ in previous definition.
- An **extended aggregation operator** is an extended function

$$A : \bigcup_{n\in\mathbb{N}} [0, 1]^n \longrightarrow [0, 1]$$

such that its every restriction to $[0, 1]^n$, for all $n \in \mathbb{N}$, is monotone in each variable, and fulfills boundary conditions

$$A(0, 0, \ldots, 0) = 0 \quad \text{and} \quad A(1, 1, \ldots, 1) = 1.$$

2.1 *Construction of Fuzzy n-Ary Aggregation Operators from a Family of Binary Aggregation Operators*

Lattice valued n-ary aggregation operators associated with binary aggregation operators have been introduced in [4]. For research in the present note, the focus is only on [0, 1]-valued binary, and n-ary aggregation operators, therefore the following forms of the results from [4] are needed.

Definition 1 [4] Let $n \geq 2$. Let $A : [0, 1]^n \longrightarrow [0, 1]$ be an n-ary function, monotone on all coordinates, satisfying $A(x, 0 \ldots, 0) = x$, and

$$\max_{i=1}^{n} x_i \leq A(x_1, \ldots, x_n).$$

Then, the family of binary operators associated with A, denoted as $\{@_i \mid i \in \{1, \ldots, n-1\}\}$, where $@_i : [0, 1]^2 \to [0, 1]$ for all $i \in \{1 \ldots n-1\}$, is of the following form

$$\begin{aligned}
@_1(x, y) &= A(x, y, 0, \ldots, 0),\\
@_2(x, y) &= A(x, y, y, \ldots, 0),\\
&\ldots\\
@_{n-1}(x, y) &= A(x, y, y, \ldots, y).
\end{aligned}$$

We can note that the obtained operators are not commutative.

Proposition 1 *[4] Let* $\{@_i \mid i \in \{1, \ldots, n-1\}\}$, $@_i : [0, 1]^2 \to [0, 1]$, $i \in \{1 \ldots n-1\}$, *be the family of functions from Definition 1. Then,* 0 *is the right neutral element for all* $@_i$ $(@_i(x, 0) = x)$, $\max\{x, y\} \leq @_i(x, y)$ *for all* $i \in \{1 \ldots n-1\}$, *all* $@_i$ *are monotone at both coordinates and the following holds*

$$@_1(x, y) \leq @_2(x, y) \leq \ldots \leq @_{n-1}(x, y).$$

Proposition 2 *[4] Let* $\{@_i \mid i \in \{1 \ldots n-1\}\}$ *be a family of monotone functions* $@_i : [0, 1]^2 \rightarrow [0, 1]$ *for all* $i \in \{1 \ldots n-1\}$ *satisfying* $@_i(x, 0) = 0$ *for all* $i \in \{1 \ldots n-1\}$, $\max\{x, y\} \leq @_i(x, y)$ *for all* i.

Let $A_@(x_1, x_2, \ldots, x_n) = @_{n-1}(@_{n-2}(\ldots @_2(@_1(x_1, x_2), x_3), \ldots, x_{n-1}), x_n)$ *be a function* $A_@ : [0, 1]^n \longrightarrow [0, 1]$. *Then* $A_@(x_1, x_2, \ldots, x_n)$ *is monotone at all coordinates and satisfies* $A_@(x, 0, \ldots, 0) = x$, *and*

$$\max_{i=1}^{n} x_i \leq A(x_1, \ldots, x_n).$$

3 Aggregation of T-Fuzzy Subgroups

The main focus of this section is on the aggregation of T-fuzzy subgroups. In general, aggregation A of two fuzzy subgroups μ and ν of a group G, is a function

$$A(\mu, \nu) : G \rightarrow [0, 1] \quad \text{defined by} \quad A(\mu, \nu)(x) = A(\mu(x), \nu(x))$$

(see [2]).

The main question is when the aggregation of two T-fuzzy subgroups is again a T-fuzzy subgroup.

In the following example, we describe T_D-fuzzy subgroups, where T_D is the drastic t-norm.

Example 1 Let $(G, *, ^{-1}, e)$ be a Klein four element group given by: $G = \{e, a, b, c\}$, $a * a = b * b = c * c = e$, $a * b = c$, $a * c = b$, $b * c = a$, and all elements are inverse to themselves.

Now, the T_D-fuzzy subgroups are all mappings of one of the three types:

	e	a	b	c
μ_1	1	p	q	q

	e	a	b	c
μ_2	1	q	p	q

	e	a	b	c
μ_3	1	q	q	p

with $0 \leq q \leq p \leq 1$.

The only non-trivial condition to check is $T_D(\mu(x), \mu(y)) \leq \mu(x * y)$.

The left-hand side is 0 and the inequality is true, except in case $\mu(x) = 1$ or $\mu(y) = 1$. In case $x = e$ or $y = e$ (e.g. $x = e$), we have $\mu(y) \leq \mu(e * y)$, which is true. Suppose that $x \neq e$ and $y \neq e$. In case $q = 1$, then all values are 1 and the inequality is fulfilled. Suppose that $q \neq 1$ and $p = 1$. Then, e.g. in case of μ_1, $T_D(\mu_1(a), \mu_1(b)) = q \leq q = \mu(c) = \mu(a * b)$. The proof in other cases is analogous. □

The following theorem is a slight generalization of Theorem 3.14 from [1]. It is said that an aggregation operator A **dominates** T if $A(T(a, b), T(c, d)) \geq T(A(a, c), A(b, d))$, for all $a, b, c, d \in [0, 1]$. In the following theorem the requirement is not that A dominate T, but only that this property is valid on the codomains of functions μ and ν. However, the proof is similar to the one in [1], so it is not included here.

Theorem 1 *Let μ, ν be T-fuzzy subgroups of a group $(G, *, ^{-1}, e)$. Let A be a binary aggregation operator satisfying*

$$A(T(a,b), T(c,d)) \geq T(A(a,c), A(b,d)),$$

for all $a, c \in \mu([0,1])$ and $b, d \in \nu([0,1])$. Then $A(\mu, \nu)$ is a T-fuzzy subgroup.

It should be noted that the condition

$$A(T(a,b), T(c,d)) \geq T(A(a,c), A(b,d)),$$

for all $a, c \in \mu([0,1])$ and $b, d \in \nu([0,1])$, is very restrictive, especially in case when A is a disjunctive aggregation operator. Namely, for A being a disjunctive aggregation operator, if $a = d = 1$, then $T(1, b) = b$ and $T(c, 1) = c$, hence $A(1, c) = A(1, d) = 1$ and $T(1, 1) = 1$. So, we would have $A(b, c) \geq 1$ which is true only if $b = 1$ or $c = 1$.

Therefore, from $\mu(x) = 1$ and $\nu(y) = 1$, it would follow that $\mu(y) = 1$ or $\nu(x) = 1$.

However, for A being a strictly monotone aggregation operator, condition

$$A(T(a,b), T(c,d)) \geq T_D(A(a,c)A(b,d))$$

is satisfied, as it is proved in the following lemma.

Lemma 1 *Let A be a strictly monotone aggregation operator, T any T-norm, and T_D drastic T norm. Then,*

$$A(T(a,b), T(c,d)) \geq T_D(A(a,c), A(b,d))$$

is satisfied for all $a, b, c, d \subset [0,1]$.

Proof For $A(a,c) \neq 1$ and $A(b,d) \neq 1$, $T_D(A(a,c), A(b,d)) = 0$ holds. In this case, the statement is valid.

Suppose that $A(a,c) = 1$. Since $A(1,1) = 1$ and in case $(a,c) \neq (1,1)$ we would have $A(a,c) \neq 1$. Hence, we got $a = 1$ and $c = 1$. Therefore, $T_D(A(a,c), A(b,d)) = A(b,d)$.

Furthermore, $A(T(a,b), T(c,d)) = A(T(1,b), (1,d)) = A(b,d)$, and the statement is proved.

Proposition 3 *Let μ, ν be T-fuzzy subgroups of a group $(G, *, ^{-1}, e)$. Let A be a strictly monotone aggregation operator. Then, $A(\mu, \nu)$ is a T_D-fuzzy subgroup.*

Proof By Lemma 1 we have that

$$A(T(a,b), T(c,d)) \geq T_D(A(a,c), A(b,d))$$

is satisfied for all $a, b, c, d \in [0, 1]$.

Hence,

$$\begin{aligned} A(\mu, \nu)(x * y) &= A(\mu(x * y), \nu(x * y)) \\ &\geq A(T(\mu(x), \mu(y)), T(\nu(x), \nu(y))) \\ &\geq T_D(A(\mu(x), \nu(x)), A(\mu(x), \nu(y))) \\ &= T_D(A(\mu, \nu)(x), A(\mu, \nu)(y)). \end{aligned}$$

In the paper [2], it has been proved that in the case of a group with two or three elements, for any T-norm, and any aggregation operation, if μ and ν are T-fuzzy subgroups, then $A(\mu, \nu)$ is a T_D-fuzzy subgroup. Proposition 3 extends the result from [2] to any group, not only for groups with 2 or 3 elements, under the assumption that A is a strict aggregation operator. Paper [2] also considers groups with at least four elements for an arbitrary (not necessarily strict) aggregation operator, however, some additional constraints, like $T = T_M$, are needed.

Extension of this problem to n-ary aggregation has been given in [1, 8]. Again, the research started from groups with 2 or 3 elements, through cyclic groups with 4 and 5 elements, and groups with prime order greater than 5, and finally finished with infinite groups taking into account the preservation of products, with some restrictions.

The result that follows addresses the problem of n-ary aggregation for any group in case the aggregation of two T-fuzzy groups is again a T-fuzzy group. In this theorem, n-ary aggregation operators are general, it is not needed that they are disjunctive as in [4].

Theorem 2 *Let $\{\mu_1, \mu_2, \ldots, \mu_n\}$ be a family of T-fuzzy subgroups of a group $(G, *, ^{-1}, e)$ with the same t-norm T.*

Let $\{A_i \mid i \in \{1 \ldots n-1\}\}$ be a family of binary aggregation operations such that $A_i(\mu, \nu)$ is a T-fuzzy subgroup, for any two T-fuzzy subgroups μ and ν.

Let $A(x_1, x_2, \ldots, x_n) = A_{n-1}(A_{n-2}(\ldots A_2(A_1(x_1, x_2), x_3), \ldots, x_{n-1}), x_n)$ be a function $A : [0, 1]^n \longrightarrow [0, 1]$ defined by the given family of binary aggregation operations.

Then $A(\mu_1, \mu_2, \ldots, \mu_n)$ is a T-fuzzy subgroup.

Proof It is easy to proof by induction that $A(\mu_1, \mu_2, \ldots, \mu_n)$ is an n-ary aggregation operator.

The proof continues by the induction on n. For $n = 2$ it follows from the assumption.

Suppose that for $\mu_1, \ldots, \mu_{n-1}$,

$$A_{n-2}(\ldots A_2(A_1(\mu_1, \mu_2), \mu_3), \ldots, \mu_{n-1})$$

is a T-fuzzy subgroup, that we denote by ϕ. Now, $A(\mu_1, \mu_2, \ldots \mu_{n-1}, \mu_n) = A_{n-1}(\phi, \mu_n)$ is a T-fuzzy subgroup by the assumption, since ϕ and μ_n are T-fuzzy subgroups.

Since the requirement that all binary aggregation operators preserve all T-fuzzy subgroups is very strong, we state a theorem with a weaker requirement, but with a similar proof which will be omitted.

Theorem 3 *Let $\mathcal{F}_0$ be a family of T-fuzzy subgroups of a group $(G, *, ^{-1}, e)$ with the same t-norm T. Let $\{A_i \mid i \in \{1 \ldots n-1\}\}$ be a family of binary aggregation operations. Let*

$$\mathcal{F}_i = \mathcal{F}_{i-1} \cup \{A_i(\mu_u, \mu_v) \mid \mu_u, \mu_v \in \mathcal{F}_{i-1}\}$$

for $i \in \{1, \ldots, n-1\}$. Let $A_i(\mu, \nu)$ be a T-fuzzy subgroup, for any two T-fuzzy subgroups $\mu, \nu \in \mathcal{F}_{i-1}$, $i \in \{1, \ldots, n-1\}$. Let

$$A(x_1, x_2, \ldots, x_n) = A_{n-1}(A_{n-2}(\ldots A_2(A_1(x_1, x_2), x_3), \ldots, x_{n-1}), x_n)$$

be a function $A : [0, 1]^n \longrightarrow [0, 1]$ defined by the given family of binary aggregation operations.

Then $A(\mu_1, \mu_2, \ldots, \mu_n)$ is a T-fuzzy subgroup for all $\mu_i \in \mathcal{F}_0$, $i \in \{1, \ldots, n\}$.

This methodology can easily be applied in all cases when the aggregation results are known in the case of binary aggregation operators. In such situations, the conclusion is transferred to the preservation of T-fuzzy subgroups under the related n-ary aggregation operator.

In the sequel, an example of aggregation of T-groups on the Klein group, which is the group of the smallest order that is not discussed in papers [1, 2, 8], is presented. The Klein group is chosen since it is not of prime order, neither is a cyclic group, and its lattice of subgroups is not chain.

Example 2 Let $(G, *, ^{-1}, e)$ be a Klein group, as in Example 1. Let us consider the following family of T_M-subgroups of this group: $\mathcal{F}_0 = \{\mu_1, \mu_2, \mu_3\}$.

	e	a	b	c
μ_1	1	1	0.7	0.7

	e	a	b	c
μ_2	1	0.9	0.3	0.3

	e	a	b	c
μ_3	1	0.7	0.6	0.6

Let A_1 be the drastic t-norm T_D, let A_2 be a special aggregation operator defined with:

$$A_2(x, y) = \begin{cases} \max\{x, y\} \text{ if } & x + y > 1 \\ \min\{x, y\} \text{ if } & x + y \le 1 \end{cases},$$

and let A_3 be the minimum t-norm T_M.

Of course, A_1 and A_3 are aggregation operators since they are t-norms. It remains to be checked whether A_2 is an aggregation operator. Obviously, $A(0, 0) = 0$ and $A(1, 1) = 1$. Suppose that $x \leq z$ and $y \leq t$. Then $\min\{x, y\} \leq \min\{y, t\}$ and $\max\{x, y\} \leq \max\{y, t\}$. In the remaining case, when $x + y \leq 1 < z + t$, $\min\{x, y\} \leq \max\{y, t\}$ holds, hence $A_2(x, y) \leq A_2(z, t)$, and the observed operator is monotone.

Now, the conditions from Theorem 3 need to be checked:

	e	a	b	c
$T_D(\mu_1, \mu_2)$	1	0.9	0	0

	e	a	b	c
$T_D(\mu_1, \mu_3)$	1	0.7	0	0

	e	a	b	c
$T_D(\mu_2, \mu_3)$	1	0	0	0

	e	a	b	c
$T_D(\mu_1, \mu_1)$	1	1	0	0

	e	a	b	c
$T_D(\mu_2, \mu_2)$	1	0	0	0

	e	a	b	c
$T_D(\mu_3, \mu_3)$	1	1	0	0

$$\mathcal{F}_1 = \{\mu_1, \mu_2, \mu_3\} \cup \{T_D(\mu_i, \mu_j) \mid i, j \in \{1, 2, 3\}\}.$$

The next step is to apply operator A_2 to all elements from $\mathcal{F}_1$. Note that $A_2(\mu, \mu) = \mu$. Moreover, if two values of one of the two T_M-fuzzy groups are 0, then aggregation A_2 always gives a T_M-fuzzy group.

The only new aggregation that is not already listed is

	e	a	b	c
$A_2(\mu_1, \mu_2)$	1	1	0.3	0.3

Finally, when operator $A_3 = T_M$ is applied to any two T_M fuzzy groups, a T_M fuzzy group is obtained. Indeed, let μ, ν be T_M-fuzzy groups. Then

$$\begin{aligned}
T_M(T_M(\mu, \nu)(x), T_M(\mu, \nu)(y)) &= T_M(T_M(\mu(x), \nu(x)), T_M(\mu(y), \nu(y))) \\
&= T_M((T_M(\mu(x), \mu(y)), T_M(\nu(x), \nu(y))) \\
&\leq T_M(\mu(x * y), \nu(x * y))' \\
&= T_M(\mu, \nu)(x * y),
\end{aligned}$$

due to associativity and monotonicity.

Hence, all the conditions of Theorem 3 are met and we can conclude that the n-ary aggregation operator applied to all T_M fuzzy groups from $\mathcal{F}_0$ gives a T_M-fuzzy group. Namely for all $\mu, \nu, \tau, \psi \in \mathcal{F}_0$,

$$A(\mu, \nu, \tau, \psi) = A_3(A_2(A_1(\mu, \nu), \tau), \psi)$$

is a T_M-fuzzy group. □

4 Conclusions and Future Work

We have developed a technique for obtaining a T-fuzzy subgroup as the result of aggregation of n T-fuzzy subgroups by n-ary aggregation operators. For this purpose, the analogous results for binary aggregation operators were used. The future work will be directed to the lattice-valued setting.

References

1. Ardanza-Trevijano, S., Chasco, M.J., Elorza, J., de Natividade, M., Talavera, F.J.: Aggregation of T-subgroups. Fuzzy Sets Syst. **463**, 108390 (2023)
2. Bejines, C., Chasco, M.J., Elorza, J.: Aggregation of fuzzy subgroups. Fuzzy Sets Syst. **418**, 170–184 (2021)
3. Beliakov, G., Pradera, A., Calvo, T.: Aggregation Functions: A Guide for Practitioners. Springer, Berlin (2007)
4. Cornejo, M.E., Medina, J., Štajner-Papuga, I., Tepavčević, A.: Lattice valued aggregation operators applicable in digital forensics. In: Computational Intelligence and Mathematics for Tackling Complex Problems 5. Studies in Computational Intelligence, vol. 1127, pp. 101–108. Springer, Cham (2024)
5. Grabisch, M., Marichal, J., Mesiar, R., Pap, E.: Aggregations Functions. Cambridge University Press (2009)
6. Karaçal, F., Mesiar, R.: Aggregation functions on bounded lattices. Int. J. Gen. Syst. **46**(1), 37–51 (2017)
7. Santos, H., Dimuro, G.P., Asmus, T.C., et al.: General Grouping Functions. Information Processing and Management of Uncertainty in Knowledge-Based Systems, vol. 1238, pp. 481–495 (2020)
8. Talavera, F.J., Ardanza-Trevijano, S., Bragard, J., Elorza, J.: Aggregation of T-subgroups of groups whose subgroup lattice is a chain. Fuzzy Sets Syst. **473**, 108717 (2023)

Independent Subcontexts in Fuzzy Formal Concept Analysis

Roberto G. Aragón, Jesús Medina, and Eloísa Ramírez-Poussa

Abstract Independent subcontext is a fundamental notion for different purposes, such as in factorization of contexts in both the Boolean and fuzzy case. Although the notion can be intuitive, a mathematical definition is required. This paper formalizes this notion and provides some properties of the independent subcontexts of an L-fuzzy context.

Keywords Formal concept analysis · Factorization · Independent subcontext

1 Introduction

Managing large tables of data and extracting useful information from them is one of the most important tasks in the modern world. One of the mathematical theories that addressess this issue is Formal Concept Analysis (FCA) which provides a mechanism to represent information through the algebraic structure of the complete lattice. Over time, extensions of this theory to more general environments such as the residuated and multi-adjoint frameworks have emerged [4, 14].

Different techniques have been developed in FCA to reduce the complexity of data processing [1, 5, 6, 8, 12]. One of these techniques is the so-called factorization [5, 11, 15] which provides a mechanism to decompose a table of data into smaller

Partially supported by the 2014–2020 ERDF Operational Programme in collaboration with the State Research Agency (AEI) in project PID2019-108991GB-I00, and with the Department of Economy, Knowledge, Business and University of the Regional Government of Andalusia in project FEDER-UCA18 108612, and by the European Cooperation in Science and Technology (COST) Action CA17124.

R. G. Aragón (✉) · J. Medina · E. Ramírez-Poussa
Department of Mathematics, University of Cádiz, Cádiz, Spain
e-mail: roberto.aragon@uca.es

J. Medina
e-mail: jesus.medina@uca.es

E. Ramírez-Poussa
e-mail: eloisa.ramirez@uca.es

L. T. Kóczy et al. (eds.), *Computational Intelligence and Mathematics for Tackling Complex Problems 6*, Studies in Computational Intelligence 1222,
https://doi.org/10.1007/978-3-031-97879-1_2

tables, significantly reducing the size and complexity of the analysis. In a classical framework, a way to find and obtain independent subcontexts of a given formal context by means of modal operators has been analyzed in [11]. Several attempts to extend this technique to a fuzzy setting have been reported [2, 13]. Recently, a study on the fuzzy extension of the methodology considered in [11], based on the fuzzy necessity operator was presented in [3]. As a consequence, it is fundamental to determine independent subcontexts in fuzzy frameworks, as well as what conditions must be satisfied in the context for the existence of this type of subcontext. However, the formal definition of independent subcontext was not introduced in these papers.

This paper mathematically formalizes this notion and goes a step further by presenting some results that establish the existence of independent subcontexts in the residuated lattice framework.

2 Preliminaries

We recall some notions related to some specific algebraic structures and FCA. The first algebraic structure that we recall is the well-known notion of horizontal sum.

Definition 1 A bounded poset $(P, \leq, \perp, \top)$ is called a *horizontal sum* of the bounded posets $\{(P_i, \leq_i, \perp, \top) \mid i \in I\}$ with I an index set, if $\bigcup_{i \in I} P_i = P$ with $P_i \cap P_j = \{\perp, \top\}$, for all $i, j \in I$, with $i \neq j$, and $x \leq y$ if and only if there is an $i \in I$ such that $\{x, y\} \subseteq L_i$ and $x \leq_i y$.

Now, let us recall the notion of complete residuated lattice.

Definition 2 A *complete residuated lattice* is an algebraic structure $(L, \leq, \perp, \top, *, \rightarrow)$ such that

1. $(L, \leq, \perp, \top)$ is a complete lattice where $\perp$ and $\top$ are the bottom and top elements, respectively.
2. $(L, *, \top)$ is a commutative monoid.
3. $(*, \rightarrow)$ is an adjoint pair, i.e., $x * y \leq z$ if and only if $x \leq y \rightarrow z$, for all $x, y, z \in L$.

In fuzzy set theory, the binary operation $*$ of a complete residuated lattice is generally referred to as a triangular norm (t-norm, for short). The well-known Gödel, product and Łukasiewicz t-norms, which are continuous t-norms on the real unit interval, are the most usual examples of t-norms.

Hereon, we will consider a complete residuated lattice $(L, \leq, \perp, \top, *, \rightarrow)$.

Definition 3 An L-*fuzzy context* is a tuple (A, B, R) such that A and B are non-empty sets (usually interpreted as attributes and objects, respectively) and R is an L-fuzzy relation $R \colon A \times B \rightarrow L$.

Next, we recall a particular type of L-fuzzy context which will be considered throughout this paper.

Definition 4 An L-fuzzy context (A, B, R) is called *normalized* if for every attribute $a \in A$ there exist $b_1, b_2 \in B$ such that $R(a, b_1) \neq \bot$ and $R(a, b_2) = \bot$ and for every object $b \in B$ there exist $a_1, a_2 \in B$ such that $R(a_1, b) \neq \bot$ and $R(a_2, b) = \bot$.

In addition, the extensions of the derivation operators $^\uparrow : L^B \to L^A$ and $^\downarrow : L^A \to L^B$ are defined as

$$g^\uparrow(a) = \inf\{g(b) \to R(a, b) \mid b \in B\}$$
$$f^\downarrow(b) = \inf\{f(a) \to R(a, b) \mid a \in A\}$$

for all $g \in L^B$, $f \in L^A$ and $a \in A$, $b \in B$, where L^B and L^A denote the set of mappings $g: B \to L$ and $f: A \to L$, respectively. The pair of concept forming operators $(^\uparrow, ^\downarrow)$ forms an antitone Galois connection between $(L^B, \leq)$ and $(L^A, \leq)$. An *L-fuzzy concept* is a pair $\langle g, f\rangle$, where $g \in L^B$ is a fuzzy subset of objects and $f \in L^A$ is a fuzzy subset of attributes, satisfying that $g^\uparrow = f$ and $g^\downarrow = g$. The set of all L-fuzzy concepts is denoted by $\mathcal{CL}$, and with the ordering defined by $\langle g_1, f_1\rangle \preceq \langle g_2, f_2\rangle$ if and only if $g_1 \leq g_2$ (equivalently $f_2 \leq f_1$), for all $\langle g_1, f_1\rangle, \langle g_2, f_2\rangle \in \mathcal{CL}$ forms a complete lattice.

Additionally, the fuzzy sets $g \in L^B$ and $f \in L^A$ such that $g(b) = \top$, for all $b \in B$, and $f(a) = \top$, for all $a \in A$, are denoted as $g_\top$ and $f_\top$, respectively. Similarly, when $g(b) = \bot$, for all $b \in B$, and $f(a) = \bot$, for all $a \in A$, they are denoted as $g_\bot$ and $f_\bot$, respectively. Moreover, we will consider a specific family of L-fuzzy subsets of L^A:

Definition 5 For each $a \in A$, the L-fuzzy subsets of attributes $\phi_{a,x} \in L^A$ are defined, for all $x \in L$, as

$$\phi_{a,x}(a') = \begin{cases} x & \text{if } a' = a \\ \bot & \text{if } a' \neq a \end{cases}$$

will be called *L-fuzzy attributes*. The set of all L-fuzzy attributes will be denoted as $\Phi = \{\phi_{a,x} \mid a \in A, x \in L\}$.

Now, we recall the characterization of the $\wedge$-irreducible elements of L-fuzzy concept lattices.

Theorem 1 ([7]) *The set of the $\wedge$-irreducible elements of $\mathcal{CL}$ is:*

$$M_F(A) = \{\langle \phi_{a,x}^\downarrow, \phi_{a,x}^{\downarrow\uparrow}\rangle \mid \phi_{a,x}^\downarrow \neq \bigwedge\{\phi_{a_i,x_i}^\downarrow \mid \phi_{a_i,x_i} \in \Phi, \phi_{a,x}^\downarrow < \phi_{a_i,x_i}^\downarrow\}; \phi_{a,x}^\downarrow \neq g_\top\}$$

3 Independent Subcontexts

This section presents an extension of the study initiated in [2]. Specifically, we formalize the notion of independent subcontexts of an L-fuzzy context and study the relationship of the original concept lattice with the attributes (and objects) of the independent subcontexts. Throughout this section, we will consider a fixed residuated lattice $(L, \leq, \bot, \top, *, \rightarrow)$ whose conjunctors $*$ has not zero-divisors and a normalized L-fuzzy context [10], that is, any object of the context is related to at least one attribute, and vice versa.

The first step to begin our study is to formalize the notion of independent subcontext. Although this notion has already been previously considered in other works as [3, 11], a formal definition has not yet been provided.

Definition 6 Let $(L, \leq, \bot, \top, *, \rightarrow)$ be a complete residuated lattice and (A, B, R) an L-fuzzy context, an *independent subcontext* is a tuple $(Y, X, R_{Y\times X})$ such that

- $Y \subset A$ and $X \subset B$ are non-empty sets.
- $R(a, b) = \bot$, for all $(a, b) \in Y^c \times X$.
- $R(a, b) = \bot$, for all $(a, b) \in Y \times X^c$.
- there exist $a \in Y$ and $b \in X$ such that $R(a, b) \neq \bot$.

The first result relates the independent subcontexts of an L-fuzzy context to the concepts of the concept lattice.

Theorem 2 *Let (A, B, R) be an L-fuzzy context, Λ an index set, and $\{A_\lambda \subset A \mid \lambda \in \Lambda\}$ and $\{B_\lambda \subset B \mid \lambda \in \Lambda\}$ two families of subsets of A and B, respectively. Then, (A, B, R) has a decomposition into independent subcontexts $\{(A_\lambda, B_\lambda, R_{A_\lambda \times B_\lambda}) \mid \lambda \in \Lambda\}$, if and only if for each concept $\langle g, f\rangle \in \mathcal{CL}$, such that $g \neq g_\top$ and $g \neq g_\bot$, there exists $\lambda \in \Lambda$, such that*

$$\langle g, f\rangle = \bigwedge\{\langle \phi^{\downarrow}_{a,x}, \phi^{\downarrow\uparrow}_{a,x}\rangle \in M_F(A) \mid a \in A_\lambda, \langle g, f\rangle \preceq \langle \phi^{\downarrow}_{a,x}, \phi^{\downarrow\uparrow}_{a,x}\rangle\}$$

and $\langle g, f\rangle \npreceq \langle \phi^{\downarrow}_{a',x}, \phi^{\downarrow\uparrow}_{a',x}\rangle$ with $\langle \phi^{\downarrow}_{a',x}, \phi^{\downarrow\uparrow}_{a',x}\rangle \in M_F(A)$, for all $a' \in A^c_\lambda$.

In particular, the previous result states that when independent subcontexts exist, each concept of the concept lattice can be obtained from $\wedge$-irreducible concepts generated from attributes that belong to the same subcontext. The following example illustrates the previous result.

Example 1 Let us consider the complete residuated lattice $(L, \leq, \bot, \top, *_G, \rightarrow_G)$ where $L = \{0, 0.25, 0.5, 0.75, 1\}$ and $*_G$ is the Gödel t-norm on L and the L-fuzzy context (A, B, R) is composed of the set of attributes $A = \{a_1, a_2, a_3\}$, the set of objects $B = \{b_1, b_2, b_3\}$ and the L-fuzzy relation R given in Table 1.

From this L-fuzzy context, it is clear that (A, B, R) can be decomposed into the independent subcontexts $(A_1, B_1, R_{A_1\times B_1})$ and $(A_2, B_2, R_{A_2\times B_2})$, where $A_1 = \{a_1, a_3\}$, $A_2 = \{a_2\}$, $B_1 = \{b_1, b_3\}$ and $B_2 = \{b_2\}$, according to Definition 6.

Therefore, as Theorem 2 states, each L-fuzzy concept, different from the top and the bottom, can be expressed as infimum of $\wedge$-irreducible concepts generated from attributes that belong to the same subcontext. For those concepts that are $\wedge$-irreducible concepts it is easy to check that they satisfy Theorem 2. This can be

Table 1 L-fuzzy relation of R Example 1

R	b_1	b_2	b_3
a_1	0.25	0	1
a_2	0	0.5	0
a_3	0.5	0	0

$$C_2 = \langle \phi^{\downarrow}_{a_2,0.75}, \phi^{\downarrow\uparrow}_{a_2,0.75} \rangle = \langle \phi^{\downarrow}_{a_2,1}, \phi^{\downarrow\uparrow}_{a_2,1} \rangle$$
$$C_3 = \langle \phi^{\downarrow}_{a_3,0.75}, \phi^{\downarrow\uparrow}_{a_3,0.75} \rangle = \langle \phi^{\downarrow}_{a_3,1}, \phi^{\downarrow\uparrow}_{a_3,1} \rangle$$
$$C_4 = \langle \phi^{\downarrow}_{a_1,0.5}, \phi^{\downarrow\uparrow}_{a,0.5} \rangle = \langle \phi^{\downarrow}_{a_1,0.75}, \phi^{\downarrow\uparrow}_{a_1,0.75} \rangle = \langle \phi^{\downarrow}_{a_1,1}, \phi^{\downarrow\uparrow}_{a_1,1} \rangle$$
$$C_5 = \langle \phi^{\downarrow}_{a_3,0.25}, \phi^{\downarrow\uparrow}_{a_3,0.25} \rangle = \langle \phi^{\downarrow}_{a_3,0.5}, \phi^{\downarrow\uparrow}_{a_3,0.5} \rangle$$
$$C_6 = \langle \phi^{\downarrow}_{a_1,0.25}, \phi^{\downarrow\uparrow}_{a_1,0.25} \rangle$$
$$C_8 = \langle \phi^{\downarrow}_{a_2,0.25}, \phi^{\downarrow\uparrow}_{a_2,0.25} \rangle = \langle \phi^{\downarrow}_{a_2,0.5}, \phi^{\downarrow\uparrow}_{a_2,0.5} \rangle$$

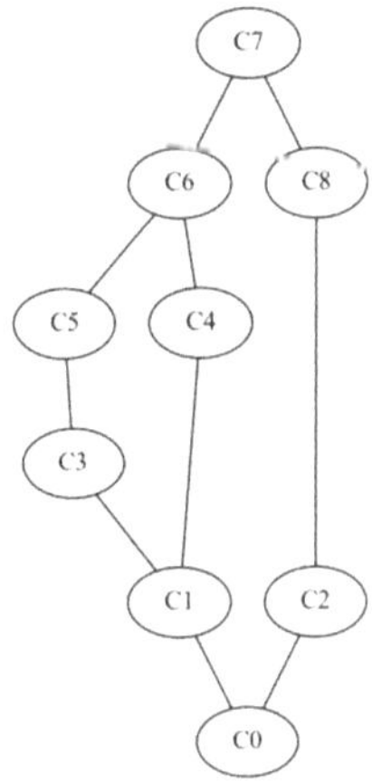

Fig. 1 $\wedge$-irreducible concepts and the L-fuzzy concept lattice of Example 1

checked on the right side of Fig. 1, where the list of $\wedge$-irreducible concepts is given, together with the fuzzy attributes that generate them. This figure also displays the L-fuzzy concept lattice associated with (A, B, R). In addition, it is easy to verify that every concept from $\{\langle \phi^{\downarrow}_{a,x}, \phi^{\downarrow\uparrow}_{a,x} \rangle \in M_F(A) \mid a \in A_1\} = \{C_3, C_4, C_5, C_6\}$ is incomparable to every concept of $\{\langle \phi^{\downarrow}_{a,x}, \phi^{\downarrow\uparrow}_{a,x} \rangle \in M_F(A) \mid a \in A_2\} = \{C_2, C_8\}$.

It remains to analyze the concept C_1, which is not a $\wedge$-irreducible concept, but it also satisfies the assumptions of Theorem 2 since this concept can be expressed as:

$$C_1 = \bigwedge \{C_3, C_4, C_5, C_6\}$$

and that $C_1 \not\preceq C$ with $C \in \{\langle \phi^{\downarrow}_{a',x}, \phi^{\downarrow\uparrow}_{a',x} \rangle \in M_F(A) \mid a' \in A_1^c = A_2\} = \{C_2, C_8\}$. □

From now on, we will consider an L-fuzzy context (A, B, R) which has independent subcontexts $\{(A_\lambda, B_\lambda, R_{A_\lambda \times B_\lambda}) \mid \lambda \in \Lambda\}$. In addition, we will denote as $\mathcal{C}(A_\lambda)$ the set of the concepts associated with the independent subcontext $(A_\lambda, B_\lambda, R_{A_\lambda \times B_\lambda})$, that is,

$$\begin{aligned} \mathcal{C}(A_\lambda) &= \mathcal{C}(A_\lambda, B_\lambda, R_{A_\lambda \times B_\lambda}) \\ &= \{\langle g, f \rangle \in \mathcal{CL} \mid \langle g, f \rangle = \bigwedge \{\langle \phi^{\downarrow}_{a,x}, \phi^{\downarrow\uparrow}_{a,x} \rangle \in M_F(A) \mid a \in A_\lambda, g \preceq \phi^{\downarrow}_{a,x}\}\} \end{aligned}$$

The following theorem analizes the algebraic structure of the set of concepts associated with an independent subcontext.

Theorem 3 *Given an L-fuzzy context (A, B, R) which has independent subcontexts $\{(A_\lambda, B_\lambda, R_{A_\lambda \times B_\lambda}) \mid \lambda \in \Lambda\}$, then $(C(A_\lambda) \cup \{\langle g_\top, f_\bot \rangle\}, \preceq_\lambda)$ is a sublattice of the original concept lattice, where $\preceq_\lambda$ is the ordering relation of the original lattice restricted to $C(A_\lambda) \cup \{\langle g_\top, f_\bot \rangle\}$.*

From the previous results and definitions, the following corollary is derived.

Corollary 1 *Given an L-fuzzy context (A, B, R) which has independent subcontexts $\{(A_\lambda, B_\lambda, R_{A_\lambda \times B_\lambda}) \mid \lambda \in \Lambda\}$, then $\bigcap_{\lambda \in \Lambda} C(A_\lambda) \subseteq \{\langle g_\bot, f_\top \rangle\}$.*

Furthermore, the concept lattices associated with the independent subcontexts considered in Theorem 3 form a horizontal sum, as the following proposition shows.

Proposition 1 *Given an L-fuzzy context (A, B, R) which has independent subcontexts $\{(A_\lambda, B_\lambda, R_{A_\lambda \times B_\lambda}) \mid \lambda \in \Lambda\}$, then $(\mathcal{CL}, \preceq)$ is a horizontal sum of the sets $\{C(A_\lambda) \cup \{\langle g_\bot, f_\top \rangle, \langle g_\top, f_\bot \rangle\} \mid \lambda \in \Lambda\}$.*

4 Conclusions and Future Work

In this paper, we have continued the study started in [3], on how to find independent subcontexts in a fuzzy environment. We have formally defined the notion of an independent subcontext in a context. In the framework of residuated lattices, we have established a relationship between concepts and independent subcontexts by giving a characterization of when a context contains independent subcontexts. Furthermore, we have shown that the concept lattice associated with a context containing independent subcontexts is a horizontal sum of sublattices.

In the near future, we will study other ways of finding independent subcontexts as well as extending these results to a more general algebraic framework such as a multi-adjoint algebra [9].

References

1. Aragón, R.G., Medina, J., Ramírez-Poussa, E.: Impact of local congruences in variable selection from datasets. J. Comput. Appl. Math. **404**, 113416 (2022)
2. Aragón, R.G., Medina, J., Ramírez-Poussa, E.: Study on the necessity operator to factorize formal contexts in a multi-adjoint framework. Commun. Comput. Inf. Sci. **1601**, 107–117 (2022)
3. Aragón, R.G., Medina, J., Ramírez-Poussa, E.: Factorizing formal contexts from closures of necessity operators. Comput. Appl. Math. **43**, 124 (2024)
4. Bělohlávek, R.: Fuzzy Galois connections. Math. Logic Q. **45**, 497–504 (1999)

5. Bělohlávek, R., Vychodil, V.: Discovery of optimal factors in binary data via a novel method of matrix decomposition. J. Comput. Syst. Sci. **76**(1), 3–20 (2010)
6. Benítez-Caballero, M.J., Medina, J., Ramírez-Poussa, E., Ślęzak, D.: Rough-set-driven approach for attribute reduction in fuzzy formal concept analysis. Fuzzy Sets Syst. **391**, 117–138 (2020)
7. Cornejo, M.E., Medina, J., Ramírez-Poussa, E.: Attribute reduction in multi-adjoint concept lattices. Inf. Sci. **294**, 41–56 (2015)
8. Cornejo, M.E., Medina, J., Ramírez-Poussa, E.: Attribute and size reduction mechanisms in multi-adjoint concept lattices. J. Comput. Appl. Math. **318**, 388–402 (2017)
9. Cornejo, M.E., Medina, J., Ramírez-Poussa, E.: Algebraic structure and characterization of adjoint triples. Fuzzy Sets Syst. **425**, 117–139 (2021)
10. Dubois, D., Medina, J., Prade, H., Ramirez-Poussa, E.: Disjunctive attribute dependencies in formal concept analysis under the epistemic view of formal contexts. Inf. Sci. **561**, 31–51 (2021)
11. Dubois, D., Prade, H.: Possibility theory and formal concept analysis: characterizing independent sub-contexts. Fuzzy Sets Syst. **196**, 4–16 (2012)
12. Dutta, S., Ślęzak, D.: Nature of decision valuations in elimination of redundant attributes. Int. J. Approx. Reason. **165**, 109091 (2024)
13. Georgescu, G., Popescu, A.: Non-dual fuzzy connections. Arch. Math. Logic **43**(8), 1009–1039 (2004)
14. Medina, J., Ojeda-Aciego, M., Ruiz-Calviño, J.: Formal concept analysis via multi-adjoint concept lattices. Fuzzy Sets Syst. **160**(2), 130–144 (2009)
15. Valverde-Albacete, F.J., Peláez-Moreno, C., Cabrera, I.P., Cordero, P., Ojeda-Aciego, M.: Formal independence analysis, pp. 596–608 (2018)

Two Sufficient Conditions for Having a Maximal Solution to Fuzzy Relational Equations

Vanja Stepanović

Abstract The existence of a maximal solution to some typical fuzzy relational equations is proved in the paper, in case when the codomain lattice is complete and meet-continuous, which means that the infimum commutes with the supremum of chains. These results extend the existing results, which are mostly limited to the existence of a maximal solution to some typical fuzzy set and fuzzy relational inequations. In order to prove that the same holds for the corresponding relational equations, it is proved that the property of meet-continuity in the codomain lattice implies another property in the lattice of fuzzy relations, namely that the composition of fuzzy relations, defined as usual, commutes with the supremum of chains. This condition can also be taken instead of the meet-continuity of the codomain lattice as another sufficient condition for the existence of a maximal solution to the equations considered here. Two examples are given, which prove that these conditions do not imply the existence of the greatest solution to some of the considered equations.

Keywords Complete lattice · Meet-continuous lattice · Fuzzy relational equations · Maximal solution

1 Introduction

Fuzzy sets were originally defined as mappings from a domain set to interval [0, 1]. Accordingly, fuzzy relations were defined as mappings from the square of the domain set to interval [0, 1]. Composition of fuzzy relations was introduced as a generalization of the composition of the usual, crisp relations [6]. Some fuzzy set and fuzzy relational equations and inequations were defined and studied using this composition (see e.g. [4]). The notions of fuzzy set, fuzzy relation and composition of fuzzy relations were generalized soon after they were introduced, so that fuzzy sets were defined as mappings from a domain set to a lattice, and fuzzy relations as mappings from the square of the domain set to the same lattice [2]. Some researchers started

---

V. Stepanović (✉)
Faculty of Agriculture, University of Belgrade, Beograd-Zemun, Serbia
e-mail: vanja@agrif.bg.ac.rs

L. T. Kóczy et al. (eds.), *Computational Intelligence and Mathematics for Tackling Complex Problems 6*, Studies in Computational Intelligence 1222,
https://doi.org/10.1007/978-3-031-97879-1_3

to investigate fuzzy set equations and inequations in this more general setting (see e.g. [3]). The sets of fuzzy sets and, particularly, fuzzy relations, are seen as lattices related to the componentwise ordering, and related to that ordering, we may speak about the greatest, the least, or maximal and minimal solutions to fuzzy relational equations.

The existence of extremal solutions to some fuzzy relational equations strongly depend on the choice of the codomain lattice. Investigations were conducted in case of Browerian and complete residuated lattices (see [3]), and also over more general algebraic structures (see [1]). The case of considering complete lattice in the codomain is the most general of these mentioned and thus some results concerning the existence and the construction of extremal solutions which were proved in case of, for example, residuated lattice (see [3]) may not hold if we take a complete lattice in the codomain. The idea is to add some conditions which will produce more results in case when the codomain lattice is complete.

For example, join-infinite distributivity of the codomain lattice implies the existence of the greatest solution to some fuzzy relational equations and inequations. Since join-infinite distributivity is a rather strong condition, taking a weaker condition of meet-continuity, we may get a proportionally weaker result. The existence of a maximal solution to some fuzzy relational inequations and two corresponding equations was proved in [5]. The statement proved here, in Lemma 2, enables us to prove the same for the rest of the corresponding equations as well.

In Preliminaries, Zorn's lemma is stated together with the few other, already known definitions. In the Results section, using Zorn's lemma, we prove the crucial Lemma 2, which enables us to prove three theorems which are sort of complementary theorems to those already proved in [5]. The Conclusion section, apart from giving a summary, contains also some ideas for future works.

What is proved in Lemma 2 is, actually, that in case of a meet-continuous codomain lattice, the set of all fuzzy relations is sort of composition-continuous, i.e. that the composition of relations commutes with the supremum of chains.

The condition of meet-continuity of the codomain lattice is not, however, sufficient for proving a stronger statement, namely the existence of the greatest solution to some typical fuzzy relational equations. It is shown here by examples.

2 Preliminaries

Lemma 1 (Zorn) *If in a non-empty poset every chain has an upper bound, there exists at least one maximal element in the poset.*

If A is a set, a fuzzy relation on A will be a mapping from $A \times A$ to a fixed complete lattice L. The set of all fuzzy relations of A will be denoted by $\mathcal{F}(A \times A)$. It is a complete lattice related to the following component-wise ordering:

$$P \leq Q \iff (\forall (x, y) \in A \times A) P(x, y) \leq Q(x, y).$$

The composition of fuzzy relations $P, Q \in \mathcal{F}(A \times A)$ will be defined as usual:

$$(P \circ Q)(x, y) = \bigvee_{z \in A} (P(x, z) \wedge Q(z, y))$$

A complete lattice L is said to be meet-continuous if the following holds:

$$a \wedge \bigvee_{i \in I} x_i = \bigvee_{i \in I} (a \wedge x_i)$$

for every $a \in L$ and every chain $\{x_i \mid i \in I\} \subseteq L$.

3 Results

First we need to show that the condition of meet-continuity in the codomain lattice implies sort of "composition-continuity" in the lattice of fuzzy relations.

Lemma 2 *If A is a set, L a complete, meet-continuous lattice, $P \in \mathcal{F}(A \times A)$ and $\{(P_i, Q_i) \mid i \in I\}$ a chain of pairs of fuzzy relations from $\mathcal{F}(A \times A)$, we have:*

$$P \circ \left(\bigvee_{i \in I} Q_i \right) = \bigvee_{i \in I} (P \circ Q_i)$$

$$\left(\bigvee_{i \in I} Q_i \right) \circ P = \bigvee_{i \in I} (Q_i \circ P)$$

$$\left(\bigvee_{i \in I} P_i \right) \circ \left(\bigvee_{i \in I} Q_i \right) = \bigvee_{i \in I} (P_i \circ Q_i)$$

Proof For all $x, y \in A$ we have:

$$\left(P \circ \left(\bigvee_{i \in I} Q_i \right) \right) (x, y) = \bigvee_{z \in A} \left(P(x, z) \wedge \left(\bigvee_{i \in I} Q_i \right) (z, y) \right)$$
$$= \bigvee_{z \in A} (P(x, z) \wedge \bigvee_{i \in I} Q_i(z, y))$$

We continue this series of equalities applying meet-continuity:

$$\begin{aligned}\bigvee_{z\in A}(P(x,z)\wedge\bigvee_{i\in I}Q_i(z,y)) &= \bigvee_{z\in A}\bigvee_{i\in I}(P(x,z)\wedge Q_i(z,y))\\ &= \bigvee_{i\in I}\bigvee_{z\in A}(P(x,z)\wedge Q_i(z,y)) = \bigvee_{i\in I}(P\circ Q_i)(x,y)\end{aligned}$$

Thus we have proved that:

$$P\circ\left(\bigvee_{i\in I}Q_i\right)=\bigvee_{i\in I}(P\circ Q_i)$$

Analogously we prove the second assertion of the lemma.

Now we prove the last assertion. Since the operation $\circ$ is monotonic with respect to both arguments, we have

$$\begin{gathered}P_i\circ Q_i\le(\bigvee_{i\in I}P_i)\circ(\bigvee_{i\in I}Q_i),\text{ thus}\\ \bigvee_{i\in I}(P_i\circ Q_i)\le(\bigvee_{i\in I}P_i)\circ(\bigvee_{i\in I}Q_i)\end{gathered}$$

Now we prove the opposite inequality:

$$\begin{gathered}(\bigvee_{i\in I}P_i)\circ(\bigvee_{i\in I}Q_i)(x,y)=\\ \bigvee_{z\in A}[(\bigvee_{i\in I}P_i)(x,z)\wedge(\bigvee_{j\in I}Q_j)(z,y)]=\\ \bigvee_{z\in A}[\bigvee_{i\in I}P_i(x,z)\wedge\bigvee_{j\in I}Q_j(z,y)]=\quad\text{(by meet-continuity)}\\ \bigvee_{z\in A}\bigvee_{j\in I}[(\bigvee_{i\in I}P_i(x,z))\wedge Q_j(z,y)]=\quad\text{(by meet-continuity)}\\ \bigvee_{z\in A}\bigvee_{j\in I}[\bigvee_{i\in I}(P_i(x,z)\wedge Q_j(z,y))]=\\ \bigvee_{z\in A}\bigvee_{(i,j)\in I^2}(P_i(x,z)\wedge Q_j(z,y))\le\\ \bigvee_{z\in A}\bigvee_{(i,j)\in I^2}[\max\{P_i,P_j\}(x,z)\wedge\max\{Q_i,Q_j\}(z,y)]=\\ \bigvee_{(i,j)\in I^2}\bigvee_{z\in A}[\max\{P_i,P_j\}(x,z)\wedge\max\{Q_i,Q_j\}(z,y)]=\\ \bigvee_{(i,j)\in I^2}[(\max\{P_i,P_j\}\circ\max\{Q_i,Q_j\})(x,y)]\le\\ \bigvee_{i\in I}(P_i\circ Q_i)(x,y)\end{gathered}$$

It was proved in [5] that for every fuzzy relation P there exists a maximal fuzzy relation contained in a given one, closed under left (right) composition with P. In terms of fuzzy relational inequations, there exists a maximal solution to the inequations $Q\circ P\le Q$ and $P\circ Q\le Q$. Using Lemma 2, we may also prove that the same holds for the corresponding equations.

Theorem 1 *Let A be a set, L a complete and meet-continuous lattice and P, Q_0, Q' fuzzy relations from $\mathcal{F}(A\times A)$ such that $Q_0\le Q'$. If $P\circ Q_0=Q_0$, there exists a solution of the equation $P\circ Q=Q$ that is maximal among those belonging to $[Q_0,Q']$.*

Proof Let $\mathcal{S}$ be the set of solutions to the equation $P \circ Q = Q$ lying in the interval $[Q_0, Q']$. It is nonempty, since $Q_0 \in \mathcal{S}$. Thus, $(\mathcal{S}, \leq)$ is a nonempty poset related to the ordering defined in $\mathcal{F}(A \times A)$.

We prove that $(\mathcal{S}, \leq)$ is closed under the supremums of the chains and thus any chain from $(\mathcal{S}, \leq)$ is bounded from above.

Let $\{Q_i \mid i \in I\}$ be a chain from $(\mathcal{S}, \leq)$. It is easily seen that its supremum lies in the interval $[Q_0, Q]$

We also prove that $Q^+ = \bigvee_{i \in I} Q_i$ is a solution to the equation $P \circ Q = Q$, using Lemma 2:

$$P \circ \bigvee_{i \in I} Q_i = \bigvee_{i \in I} P \circ Q_i = \bigvee_{i \in I} Q_i$$

Thus, $\bigvee\limits_{i \in I} Q_i \in \mathcal{S}$, and applying Zorn's lemma we prove the first assertion of the theorem. Analogously we prove the same for the equation $Q \circ P = Q$.

Since $P \circ 0 = 0$, taking $Q_0 = 0$ in Theorem 1, we get the following.

Corollary 1 *There exists a solution to the equation $P \circ Q = Q$ that is maximal among those contained in a given fuzzy relation $Q' \in \mathcal{F}(A \times A)$.*

Taking $Q' = 1$ in this corollary, we get another one.

Corollary 2 *There exists a maximal solution to the equation $P \circ Q = Q$.*

In [5] it was proved that there exists maximal solutions to the inequations $P \circ Q \leq R$ and $Q \circ P \leq R$ where P and R are given fuzzy relations and Q is an unknown relation. The same was proved for the corresponding equations $P \circ Q = R$ and $Q \circ P = R$. Also, it was proved that there exists a maximal solution—related to the componentwise ordering in $(\mathcal{F}(A \times A))^2$—to the inequation $P \circ Q \leq R$, where both P and Q were unknown, i.e. (P, Q) are an unknown pair of fuzzy relations, and (P_0, Q_0) is a solution to $P \circ Q = R$ if $P_0 \circ Q_0 = R$. But using the same methods as in [5] we could not prove that this holds for the corresponding equation $P \circ Q = R$. Now, using Lema 2, we can prove it.

Theorem 2 *Let A be a set, L a complete and meet-continuous lattice and P_0, Q_0, R relations from $\mathcal{F}(A \times A)$, such that $P_0 \circ Q_0 = R$. There exists a maximal solution to the equation $P \circ Q = R$—where P and Q are unknown—greater or equal to (P_0, Q_0). In particular, there exists a maximal solution to the equation $P \circ Q = R$.*

Proof Let $\mathcal{S}$ be the set of solutions to the equation $P \circ Q = Q$ which are greater or equal to (P_0, Q_0). It is non-empty, since $(P_0, Q_0) \in \mathcal{S}$.

If $\{(P_i, Q_i) \mid i \in I\}$ is a chain from $(\mathcal{S}, \leq)$, $\bigvee\limits_{i \in I}(P_i, Q_i) = (\bigvee\limits_{i \in I} P_i, \bigvee\limits_{i \in I} Q_i)$ and $(\bigvee\limits_{i \in I} P_i, \bigvee\limits_{i \in I} Q_i)$ is greater or equal to (P_0, Q_0).

Using the last assertion of Lemma 2, we prove that $(\bigvee\limits_{i \in I} P_i, \bigvee\limits_{i \in I} Q_i)$ is a solution to the equation $P \circ Q = R$:

$$\bigvee_{i\in I} P_i \circ \bigvee_{i\in I} Q_i = \bigvee_{i\in I} P_i \circ Q_i = \bigvee_{i\in I} R = R.$$

Thus $\bigvee_{i\in I}(P_i, Q_i) \in \mathcal{S}$ and the theorem follows by Zorn's Lemma.

In [5] it was proved that there exists a maximal transitive relation contained in a given arbitrary relation. In terms of fuzzy relational inequations, this means that there exists a maximal solution to the inequation $Q \circ Q \leq Q$ contained in a given fuzzy relation. Using Lemma 2, we may prove an analogous result for the corresponding fuzzy relational equation $Q \circ Q = Q$.

Theorem 3 *Let A be a set, L a complete and meet-continuous lattice and* $Q_0, Q' \in \mathcal{F}(A \times A)$ *such that* $Q_0 \leq Q'$ *and* $Q_0 \circ Q_0 = Q_0$. *There exists a solution of the equation* $Q \circ Q = Q$ *that is maximal among those belonging to* $[Q_0, Q']$.

Proof The proof follows from the last statement of Lemma 2, and it is analogous to that of Theorem 2.

Corollary 3 *There exists a solution to the equation* $Q \circ Q = Q$ *that is maximal among those contained in a given fuzzy relation* Q'.

A fuzzy relation is a solution to the equation $Q \circ Q = Q$ if and only if it is idempotent related to the composition of relations, thus there exists a maximal idempotent fuzzy relation contained in a given one.

Thus we have proved the existence of a maximal solution to some typical fuzzy set equations. The existence of a maximal solution to the equations $P \circ Q = R$ and $Q \circ P = R$ (where Q is the unknown)—which was proved in [5]—could also be effectively proved using Lemma 2. When these equations are concerned, we can easily see that the condition of being meet-continuous is not sufficient for having the greatest solution.

The following diamond lattice L in Fig. 1 is meet-continuous, since it is finite. Let it be the codomain lattice and $A = \{t\}$ the domain set. If P, R, Q' and Q'' are fuzzy relations defined by

$$P(t,t) = a;\ R(t,t) = 0;\ Q'(t,t) = b;\ Q''(t,t) = c.$$

Both Q' and Q'' are solutions to the equations $P \circ Q = R$ and $Q \circ P = R$. If there exists the greatest solution to these equation, it must contain both Q' and Q'', thus it must be equal to the greatest fuzzy relation T, defined by $T(t,t) = 1$. But $P \circ T = T \circ P = P \neq R$. Therefore the greatest solution to the equations $P \circ Q = R$ and $Q \circ P = R$ does not exist.

Using the same lattice and a two-element domain set $A = \{t, u\}$, we prove also that taking a complete and meet-continuous lattice in the codomain is also not sufficient for having the greatest solution to the equation $T \circ T = T$ contained in a given fuzzy relation, and also for having the greatest transitive fuzzy relation contained in a given fuzzy relation. Let P and R be defined by

Fig. 1 Lattice L

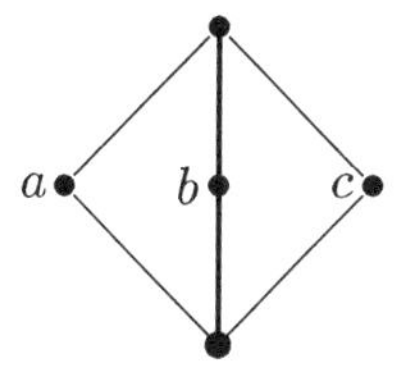

$$P(x,x) = a;\ P(x,y) = b;\ P(y,x) = c;\ P(y,y) = 1$$
$$R(x,x) = a;\ R(x,y) = c;\ R(y,x) = b;\ R(y,y) = 1.$$

Note that $P \circ P = P$ and $R \circ R = R$; if there exists the greatest solution to the equation $X \circ X = X$ (or to $X \circ X \leq X$ as well) contained in $S = P \vee R$, it contains both P and R, thus it contains S, and is equal to it. But

$$(S \circ S)(x,x) = 1 > S(x,x) = a$$

S is neither idempotent nor transitive, and the contradiction we got proves that there might not be the greatest idempotent relation contained in a given one, and also the greatest transitive relation contained in a given one might not exist.

4 Conclusion

The existence of a maximal solution to some typical fuzzy relational equations was proved in the paper. It was proved using the condition of "composition-continuity" in the lattice of fuzzy relations, proved to hold in case the codomain lattice is meet-continuous.

Since all the above theorems follow solely from the composition-continuity, that condition may independently be taken as sufficient for having a maximal solution to the considered equations.

It was, however, proved, that these conditions are insufficient for proving the existence of the greatest solution of some typical fuzzy relational equations.

In the future, the existence of maximal solutions to some equations other than these contained in this paper will be tested. An effort will be put to get some more general results in this area. Possible constructions of maximal and/or other types of extremal solutions in this setting is also a topic to investigate.

Acknowledgements This research was supported by the Science Fund of the Republic of Serbia, Grant no 6565, Advanced Techniques of Mathematical Aggregation and Approximative Equations Solving in Digital Operational Research—AT-MATADOR. It is also a result of the Agreement on the transfer of funds for the financing of scientific research work of teaching staff at accredited higher education institutions in Serbia in 2024, grant number 451-03-65/2024-03/200116.

Disclosure of Interests The author has no competing interests to declare that are relevant to the content of this article.

References

1. Díaz-Moreno, J.C., Medina, J., Turunen, E.: Minimal solutions of general fuzzy relation equations on linear carriers. An algebraic characterization. Fuzzy Sets Syst. **311**, 112–123 (2017)
2. Goguen, J.A.: L-fuzzy sets. J. Math. Anal. Appl. **18**, 145–174 (1967)
3. Ignjatović, J., Ćirić, M., Šešelja, B., Tepavčević, A.: Fuzzy relational inequalities and equalities, fuzzy quasi-orders, closures and openings of fuzzy sets. Fuzzy Sets Syst. **260**, 1–24 (2015)
4. Sanchez, E.: Resolution of composite fuzzy relation equations. Inf. Control **30**(1), 38–48 (1976)
5. Stepanović, V.: Fuzzy set inequations and equations with a meet-continuous codomain lattice. J. Intell. Fuzzy Syst. **34**(6), 4009–4021 (2018)
6. Zadeh, L.A.: Fuzzy sets. Inf. Control **8**, 338–353 (1965)

Towards Compatible Subcontext by Means of Classical Attribute Reduction

M. José Benítez-Caballero and Jesús Medina

Abstract Reducing a database is one of the main goals of the theories developed for the management of information in databases, such as Formal Concept Analysis (FCA) and Rough Set Theory (RST). On the other hand, this reduction should preserve some properties from the original database. In FCA, for example, it is recommendable to obtain compatible subcontexts. This paper introduces sufficient conditions to obtain compatible subcontexts from attribute reduction theory in FCA.

Keywords Formal concept analysis · Attribute reduction · Reduct · Compatible subcontext

1 Introduction

Last decades, several theories have been developed in order to manage information stored in databases. Two of the most powerful tolls are Formal Concept Analysis (FCA) [17] and Rough Set Theory (RST) [16], both arose in the eighties. Nevertheless, the philosophy underlining the theories are not the same, there are similarities between them. Such as the way of mathematically represents databases are quite similar: a set of objects to study over a set of properties and the values taken for each one. Also, the notion of reduct and the mechanism to obtain them are an important

Partially supported by the projects PID2019-108991GB-I00 and PID2022-137620NB-I00/AEI/10.13039/501100011033/FEDER, UE, and with grant TED2021-129748B-I00 funded by MCIN/AEI/10.13039/501100011033 and, as appropriate, by "ERDF A way of making Europe", by the 'European Union" or by the "European Union NextGenerationEU/PRTR".

M. J. Benítez-Caballero (✉) · J. Medina
Department of Mathematics, University of Cádiz, Cadiz, Spain
e-mail: mariajose.benitez@uca.es
URL: https://mcis.uca.es/

J. Medina
e-mail: jesus.medina@uca.es

L. T. Kóczy et al. (eds.), *Computational Intelligence and Mathematics for Tackling Complex Problems 6*, Studies in Computational Intelligence 1222,
https://doi.org/10.1007/978-3-031-97879-1_4

goal in both theories [2, 5, 9, 12, 13]. That is, the computation of a minimal set of attributes keeping the original information of the database.

Other reduction procedures are proposed in the literature [1, 7, 8, 14]. For example, attribute selection philosophy of RST was used to reduce concept lattices in [3, 4]. Many of these procedure do not preserve the concept lattice structure of the original context, but other weaker properties although maybe enough in specific applications.

On the other hand, the notion of compatible subcontext was introduced in [13]. This kind of contexts has interesting properties that a reduction should preserve. However, this notion has not been intensively studied in the literature. This paper studies sufficient conditions a reduction should have in order to obtain compatible subcontexts.

2 Preliminaries

FCA requires the translation of the considered dataset to a table, where a set of attributes or properties $\mathcal{P}$, a set of objects $\mathcal{O}$, both of them non empty, and a crisp relationship between them $\mathcal{R}\colon \mathcal{O} \times \mathcal{P} \to \{0, 1\}$ are considered. We said that an attribute $a \in \mathcal{P}$ and an object $b \in \mathcal{O}$ are related when the relation is defined as $\mathcal{R}(b, a) = 1$; otherwise it is defined as $\mathcal{R}(b, a) = 0$. Also, if $\mathcal{R}(b, a) = 1$, the relation can be wrote as $b\mathcal{R}a$.

A *context* is the triple $(\mathcal{O}, \mathcal{P}, \mathcal{R})$ and we can define the mappings ${}^{\uparrow}\colon 2^{\mathcal{O}} \to 2^{\mathcal{P}}$, ${}^{\downarrow}\colon 2^{\mathcal{P}} \to 2^{\mathcal{O}}$, for each $Y \subseteq \mathcal{O}$ and $A \subseteq \mathcal{P}$, as follows:

$$Y^{\uparrow} = \{a \in \mathcal{P} \mid \text{for all } b \in Y, b\mathcal{R}a\} = \{a \in \mathcal{P} \mid \text{if } b \in Y, \text{ then } b\mathcal{R}a\} \tag{1}$$
$$A^{\downarrow} = \{b \in \mathcal{O} \mid \text{for all } a \in A, b\mathcal{R}a\} = \{b \in \mathcal{O} \mid \text{if } a \in A, \text{ then } b\mathcal{R}a\} \tag{2}$$

Originally these operators were denoted as $'$ by Ganter and Wille and they were called derivation operators. In order to differentiate between the mapping on the set of objects and on the set of attributes, we have changed the notation.

A *concept* in the context $(\mathcal{O}, \mathcal{P}, \mathcal{R})$ is a pair (Y, A), where $Y \subseteq \mathcal{O}$, $A \subseteq \mathcal{P}$, $Y^{\uparrow} = A$ and $A^{\downarrow} = Y$ hold. The subset of objects Y of the concept (Y, A) is called *extent* and the *intent* is the subset of attributes A.

The set of all the concepts is denoted as $\mathcal{C}(\mathcal{O}, \mathcal{P}, \mathcal{R})$. This set has a complete lattice structure [10, 13], when we consider the inclusion ordering on the left argument (or, equivalently, the opposite of the inclusion ordering on the right argument). That means, for each $(Y_1, A_1), (Y_2, A_2) \in \mathcal{C}(\mathcal{O}, \mathcal{P}, \mathcal{R})$, we have that $(Y_1, A_1) \leq (Y_2, A_2)$, if $Y_1 \subseteq Y_2$ (or, equivalently, $A_2 \subseteq A_1$). We define the meet ($\wedge$) and join ($\vee$) operators by:

$$(Y_1, A_1) \wedge (Y_2, A_2) = (Y_1 \wedge Y_2, (A_1 \vee A_2)^{\downarrow\uparrow})$$
$$(Y_1, A_1) \vee (Y_2, A_2) = ((Y_1 \vee Y_2)^{\uparrow\downarrow}, A_1 \wedge A_2)$$

for all $(Y_1, A_1), (Y_2, A_2) \in \mathcal{C}(\mathcal{O}, \mathcal{P}, \mathcal{R})$.

Note that the operators defined in Eqs. (1) and (2) form a Galois connection. Given an attribute $a \in \mathcal{P}$, *the concept generated by* a, that is $(a^{\downarrow}, a^{\downarrow\uparrow})$, will be called *attribute-concept*. Note that this pair is really a concept due to $(^{\uparrow}, ^{\downarrow})$ is a Galois connection [10, 11]. Similarly, given an object $b \in \mathcal{O}$, the concept generated by b, that is $(b^{\uparrow\downarrow}, b^{\uparrow})$, will be called *object-concept*.

On the other hand, reducing the set of attributes without modifying the structure of the concept lattice is persued by attribute reduction theory in formal concept analysis. That means, obtaining a new concept lattice isomorphic to the original one. Now, we recall some necessary definitions related to attribute reduction in FCA.

We said two concept lattices $\mathcal{C}(\mathcal{O}, \mathcal{P}_1, \mathcal{R}_1), \mathcal{C}(\mathcal{O}, \mathcal{P}_2, \mathcal{R}_2)$ are *isomorphic* if $\mathcal{C}(\mathcal{O}, \mathcal{P}_1, \mathcal{R}_1) \leq \mathcal{C}(\mathcal{O}, \mathcal{P}_2, \mathcal{R}_2)$ and $\mathcal{C}(\mathcal{O}, \mathcal{P}_2, \mathcal{R}_2) \leq \mathcal{C}(\mathcal{O}, \mathcal{P}_1, \mathcal{R}_1)$ hold; and we will write

$$\mathcal{C}(\mathcal{O}, \mathcal{P}_1, \mathcal{R}_1) \cong \mathcal{C}(\mathcal{O}, \mathcal{P}_2, \mathcal{R}_2)$$

If we consider a context $(\mathcal{O}, \mathcal{P}, \mathcal{R})$, a subset of attributes, $A \subseteq \mathcal{P}$ and the restricted relation $\mathcal{R}_{|A} = \mathcal{R} \cap (A \times \mathcal{O})$ then, the triple $(\mathcal{O}, A, \mathcal{R}_{|A})$ is also a formal context. We can interpret it as a *subcontext* of the original one. Therefore, the mappings $^{\downarrow}$ and $^{\uparrow}$ introduced in Eqs. (1) and (2) can be defined in this subcontext they will be denoted as $^{\downarrow^A}$ and $^{\uparrow_A}$ in order to avoid some confusion. It is clear that, given $Y \subseteq \mathcal{O}$, we obtain that $Y^{\uparrow_A} = Y^{\uparrow} \cap A$. Note that, when we consider a subset $A_1 \subseteq A$, then $A_1^{\downarrow^Y} = A_1^{\downarrow}$.

Proposition 1 [3] *Given a context* $(\mathcal{O}, \mathcal{P}, \mathcal{R})$ *and a subset* $D \subseteq \mathcal{P}$. *The set* $R_E = \{((Y_1, A_1), (Y_2, A_2)) \mid (Y_1, A_1), (Y_2, A_2) \in \mathcal{C}(\mathcal{O}, \mathcal{P}, \mathcal{R}), Y_1^{\uparrow_D\downarrow} = Y_2^{\uparrow_D\downarrow}\}$ *is an equivalence relation. The classes of the quotient set* $\mathcal{C}(A, B, R)/R_E$ *will be denoted as* $[(Y, A)]_D$, *for all* $(Y, A) \in \mathcal{C}(\mathcal{O}, \mathcal{P}, \mathcal{R})$.

Considering a formal context $(\mathcal{O}, \mathcal{P}, \mathcal{R})$, it is easy to verify that for any $A \subseteq \mathcal{P}$, such that $A \neq \varnothing$, $\mathcal{C}(\mathcal{O}, \mathcal{P}, \mathcal{R}) \leq \mathcal{C}(\mathcal{O}, A, \mathcal{R}_{|A})$ holds.

Definition 1 Let $(\mathcal{O}, \mathcal{P}, \mathcal{R})$ be a context, if there exists a set of attributes $D \subseteq \mathcal{P}$ such that $\mathcal{C}(\mathcal{O}, \mathcal{P}, \mathcal{R}) \cong \mathcal{C}(\mathcal{O}, A, \mathcal{R}_{|A})$, then D is called a *consistent* set of $(\mathcal{O}, \mathcal{P}, \mathcal{R})$. Moreover, if $\mathcal{C}(\mathcal{O}, D \setminus \{a\}, \mathcal{R}_{|D\setminus\{a\}}) \not\cong \mathcal{C}(\mathcal{O}, \mathcal{P}, \mathcal{R})$, for all $a \in D$, then D is called *reduct* of $(\mathcal{O}, \mathcal{P}, \mathcal{R})$.

The *core* of $(\mathcal{O}, \mathcal{P}, \mathcal{R})$ is the intersection of all the reducts of $(\mathcal{O}, \mathcal{P}, \mathcal{R})$, which will be denoted as $C_{\mathcal{P}}$.

Analogously, these results can be written with respect to the object, obtaining an object reduction procedure. For example, the notion of reduct can be obtained as the minimal set of objects preserving the original concept lattice (by isomorphisms). In this case, we will add the letter associated with the objects, that is, we will call it $\mathcal{O}$-reduct of the context $(\mathcal{O}, \mathcal{P}, \mathcal{R})$. Moreover, a reduct $D \subseteq \mathcal{P}$ of $(\mathcal{O}, \mathcal{P}, \mathcal{R})$ will be called as $\mathcal{P}$-reduct. The core of objects of the context $(\mathcal{O}, \mathcal{P}, \mathcal{R})$ will be denoted as $C_{\mathcal{O}}$.

Next, the notion of compatible subcontext given in [13] is recalled.

Definition 2 Given a context $(\mathcal{O}, \mathcal{P}, \mathcal{R})$, the subsets $Z \subseteq \mathcal{O}$ and $D \subseteq \mathcal{P}$, the subcontext $(Z, D, R_{|Z\times D})$ is called *compatible* if the pair $(Y \cap Z, A \cap D)$ is a concept of the subcontext for every concept $(Y, A) \in \mathcal{C}(\mathcal{O}, \mathcal{P}, \mathcal{R})$.

Finally, we will recall the notions of meet-irreducible and join-irreducible elements of a lattice, since these definitions will also be considered later.

Definition 3 Given a lattice $(L, \preceq)$, such that $\wedge$, $\vee$ are the meet and the join operators, and an element $x \in L$ verifying

1. If L has a top element $\top$, then $x \neq \top$.
2. If $x = y \wedge z$, then $x = y$ or $x = z$, for all $y, z \in L$.

we call x *meet-irreducible* ($\wedge$*-irreducible*) *element* of L. A *join-irreducible* ($\vee$*-irreducible*) *element* of L is defined dually.

Given a context $(\mathcal{O}, \mathcal{P}, \mathcal{R})$, the set of join-irreducible (meet-irreducible) elements of the concept lattice will be denoted as: $J_F(\mathcal{O}, \mathcal{P}, \mathcal{R})$ $(M_F(\mathcal{O}, \mathcal{P}, \mathcal{R}))$.

3 Attribute Reduction Versus Compatible Subcontexts

This section will study different sufficient conditions on the considered reduction in order to obtain compatible subcontexts. The first results show that compatible subcontexts arise when $\mathcal{P}$-consistent sets or $\mathcal{O}$-consistent sets are considered in the reduction of the original context.

Theorem 1 *Given a $\mathcal{P}$-consistent set $D \subseteq \mathcal{P}$ of the context $(\mathcal{O}, \mathcal{P}, \mathcal{R})$, the reduced context $(\mathcal{O}, D, \mathcal{R}_{\mathcal{O}\times D})$ is a compatible subcontext.*

A similar result arises when the reduction is done in the set of objects.

Theorem 2 *Let $X \subseteq \mathcal{O}$ be a $\mathcal{O}$-consistent set of the context $(\mathcal{O}, \mathcal{P}, \mathcal{R})$, then the subcontext $(X, \mathcal{P}, \mathcal{R}_{X\times\mathcal{P}})$ is compatible.*

Finally, we obtain that the consideration of both consistent sets, in the attributes and in the objects, offer compatible subcontexts.

Corollary 1 *Given a $\mathcal{P}$-consistent set $D \subseteq \mathcal{P}$ and a $\mathcal{O}$-consistent set $X \subseteq \mathcal{O}$ of the context $(\mathcal{O}, \mathcal{P}, \mathcal{R})$, the reduced context $(X, D, \mathcal{R}_{\mathcal{O}\times D})$ is a compatible subcontext.*

Given a subset of attributes $D \subseteq \mathcal{P}$, if it is not consistent, then the concept lattice of the reduced context $(\mathcal{O}, D, \mathcal{R}_{\mathcal{O}\times D})$ is not isomorphic to the original one, and so the extents are not preserved. This is a problem with the definition of compatibility since the ordering in the concepts is increasing in the objects and decreasing in the attributes, however both sets are reduced in the definition of compatible subcontext $((Y \cap X, A \cap D)$ must be a concept of the subcontext for every concept $(Y, A) \in \mathcal{C}(\mathcal{O}, \mathcal{P}, \mathcal{R}))$.

Now, we will introduce a first advance on this line, trying to reduce this drawback from the results in [3]. Propositions 3.8 and 3.9 in the aforementioned paper show that an equivalence relation arises where the classes are join semilattices with maximum elements $(X^{\uparrow_D\downarrow}, X^{\uparrow_D\downarrow\uparrow})$, for all $X \subseteq \mathcal{O}$.

The elements in the class $[X, B]_D \subseteq \mathcal{C}(\mathcal{O}, \mathcal{P}, \mathcal{R})$ merge their extents to the extent $X^{\uparrow_D\downarrow}$ in the reduced concept lattice. Hence, by the definition of compatible subcontext, this fact must naturally be avoided. For that, the following set is considered:

$$S_D = \{x \in \mathcal{O} \mid [(x^{\uparrow_D\downarrow}, x^{\uparrow_D\downarrow\uparrow})]_D \text{ is a singleton}\}$$

Notice that, this set only considers object concepts of the reduced concept lattice $(\mathcal{O}, D, \mathcal{R}_{\mathcal{O}\times D})$. Since they include the join-irreducible elements [6], this information is enough to represent the reduced concept lattice. This set is a proper subset of the set of objects, if D contains at least an attribute in the core.

Proposition 2 *If D is a $\mathcal{P}$-consistent set, then $S_D = \mathcal{O}$.*

In this case, we need to consider another set Z_D, which is defined as the complementary set of:

$$\overline{Z_D} = \{x \in \mathcal{O} \mid x^{\uparrow\downarrow} = x^{\uparrow_D\downarrow}; [(x^{\uparrow_D\downarrow}, x^{\uparrow_D\downarrow\uparrow})]_D \text{ is not a singleton}\}$$

that is, $Z_D = \mathcal{O} \setminus \overline{Z_D}$. From this set, we obtain the following result.

Theorem 3 *Given a subset $D \subseteq \mathcal{P}$ of the context $(\mathcal{O}, \mathcal{P}, \mathcal{R})$, $a \in \mathcal{P} \setminus D$, $x \in \mathcal{O} \setminus Z_D$, if $a \in C_\mathcal{P}$, $x \in C_\mathcal{O}$, and $(a^\downarrow, a^{\downarrow\uparrow})$ is the lower neighbor of $(x^{\uparrow\downarrow}, x^\uparrow)$, then $(Z_D, D, \mathcal{R}_{Z_D\times D})$ is a compatible subcontext.*

This result can be rewritten in terms of meet and join irreducible elements.

Corollary 2 *Given a subset $D \subseteq \mathcal{P}$ of the context $(\mathcal{O}, \mathcal{P}, \mathcal{R})$, such that the cardinal of $M_F(\mathcal{O}, \mathcal{P}, \mathcal{R}) \setminus M_F(\mathcal{O}, D, \mathcal{R})$ is one and, also, the cardinal of $J_F(\mathcal{O}, \mathcal{P}, \mathcal{R}) \setminus J_F(Z_D, \mathcal{P}, \mathcal{R})$ is one, and the former is lower neighbor of the latest, then the subcontext $(Z_D, D, \mathcal{R}_{Z_D\times D})$ is compatible.*

Next, an illustrative example is presented.

Example 1 Given the context $(\mathcal{O}, \mathcal{P}, \mathcal{R})$ with three objects and four attributes related by the relation given in Fig. 1, the concept lattice is composed of five concepts.

Clearly, b and c are attributes of the core, and a and d are relatively necessary attributes [15]. Therefore, $D = \{a, b, c\}$ is a reduct of the concepts and, by Theorem 1, $(\mathcal{O}, D, \mathcal{R}_{\mathcal{O}\times D})$ is a compatible subcontext, as it can easily be checked. Now, if we remove an element in the core, we do not obtain a reduct and we cannot apply the same result. For example, if we consider the set $E = \{a, c\}$, then $[(x_3^{\uparrow_D\downarrow}, x_3^{\uparrow_D\downarrow\uparrow})]_D$ is not a singleton, since it contains to C_2 and C_3. As a consequence, we have that $\overline{Z_E} = \{x_3\}$, and so $Z_E = \{x_1, x_2\}$. Moreover, x_3 generates a join-irreducible concept, which is the upper neighbor of the meet-irreducible concept generated by b. Thus, by

R	a	b	c	d
x_1	1	0	0	1
x_2	0	1	1	0
x_3	0	0	1	0

$$C_0 = (\varnothing, \mathcal{P})$$
$$C_1 = (\{x_1\}, \{a, d\})$$
$$C_2 = (\{x_2\}, \{b, c\})$$
$$C_3 = (\{x_2, x_3\}, \{c\})$$
$$C_4 = (\mathcal{O}, \varnothing)$$

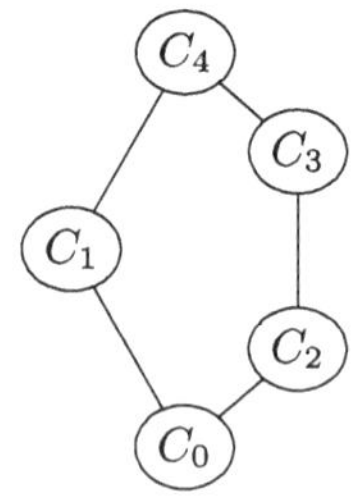

Fig. 1 Relation R, concepts and concept lattice of the context in Example 1

Theorem 3, $(Z_E, E, \mathcal{R}_{Z_E \times D})$ is a compatible subcontext. For example, we can see that, given the concepts $(\{x_2\}, \{b, c\})$ and $(\{x_2, x_3\}, \{c\})$, we obtain $(\{x_2\} \cap Z_E, \{b\} \cap E) = (\{x_2\}, \{c\})$, and $(\{x_2, x_3\} \cap Z_E, \{c\} \cap E) = (\{x_2\}, \{c\})$, which are concepts of the reduced context $(Z_E, E, \mathcal{R}_{Z_E \times D})$. Furthermore, we have that $M_F(\mathcal{O}, \mathcal{P}, \mathcal{R}) \setminus M_F(\mathcal{O}, E, \mathcal{R})$ and $J_F(\mathcal{O}, \mathcal{P}, \mathcal{R}) \setminus J_F(Z_E, \mathcal{P}, \mathcal{R})$ are singleton satisfying the required relation.

4 Conclusions and Future Work

This paper has presented sufficient conditions to ensure that an attribute and object reduction provides compatible subcontexts. This is a first step in order to introduce new properties to characterize the interesting notion of compatibility of contexts, complementing the ones given in [13]. We have shown that the usual attribute reduction in FCA, preserving the concept lattice structure of the original context, brings compatible subcontexts. However, we must take care when some meet-irreducible concept "disappears" after the reduction. This paper establishes that a compatible subcontext is obtained when only one meet-irreducible concept is removed. In the future, we will study more sufficient conditions to characterize what happens when more than one of these concepts are eliminated by the reduction. Moreover, we will study particular attribute reduction mechanisms, such that, the one given by the relationship between FCA and RST [3, 4].

References

1. Aragón, R.G., Medina, J., Ramírez-Poussa, E.: Impact of local congruences in variable selection from datasets. J. Comput. Appl. Math. 113416 (2022). https://doi.org/10.1016/j.cam.2021.113416
2. Benítez-Caballero, M.J., Medina, J., Ramírez-Poussa, E., Ślęzak, D.: Bireducts with tolerance relations. Inf. Sci. **435**, 26–39 (2018)
3. Benítez-Caballero, M.J., Medina, J., Ramírez-Poussa, E., Ślęzak, D.: A computational procedure for variable selection preserving different initial conditions. Int. J. Comput. Math. **97**(1–2), 387–404 (2020)

4. Benítez-Caballero, M.J., Medina, J., Ramírez-Poussa, E., Ślęzak, D.: Rough-set-driven approach for attribute reduction in fuzzy formal concept analysis. Fuzzy Sets Syst. **391**, 117–138 (2020)
5. Chu, X., Sun, B., Chu, X., Wu, J., Han, K., Zhang, Y., Huang, Q.: Multi-granularity dominance rough concept attribute reduction over hybrid information systems and its application in clinical decision-making. Inf. Sci. **597**, 274–299 (2022)
6. Cornejo, M.E., Medina, J., Ramírez-Poussa, E.: Attribute reduction in multi-adjoint concept lattices. Inf. Sci. **294**, 41–56 (2015)
7. Cornejo, M.E., Medina, J., Ramírez-Poussa, E.: On the use of thresholds in multi-adjoint concept lattices. Int. J. Comput. Math. **92**(9), 1855–1873 (2015)
8. Cornejo, M.E., Medina, J., Ramírez-Poussa, E.: Attribute and size reduction mechanisms in multi-adjoint concept lattices. J. Comput. Appl. Math. **318**, 388–402 (2017)
9. Cornelis, C., Medina, J., Verbiest, N.: Multi-adjoint fuzzy rough sets: definition, properties and attribute selection. Int. J. Approximate Reasoning **55**, 412–426 (2014)
10. Davey, B., Priestley, H.: Introduction to Lattices and Order, 2nd edn. Cambridge University Press (2002)
11. Denecke, K., Ern, M., Wismath, S.L. (eds.): Galois Connections and Applications. Kluwer Academic Publishers, Dordrecht. The Netherlands (2004)
12. Dutta, S., Ślęzak, D.: Nature of decision valuations in elimination of redundant attributes. Int. J. Approximate Reasoning **165**, 109091 (2024)
13. Ganter, B., Wille, R.: Formal Concept Analysis: Mathematical Foundation. Springer (1999)
14. Liu, G., Xie, Y., Gao, X.: Three-way reduction for formal decision contexts. Inf. Sci. **615**, 39–57 (2022)
15. Medina, J.: Relating attribute reduction in formal, object-oriented and property-oriented concept lattices. Comput. Math. Appl. **64**(6), 1992–2002 (2012)
16. Pawlak, Z.: Rough sets. Int. J. Comput. Inf. Sci. **11**, 341–356 (1982)
17. Wille, R.: Restructuring lattice theory: an approach based on hierarchies of concepts. In: Rival, I. (ed.) Ordered Sets, pp. 445–470. Reidel (1982)

Extension of the Time Dependent Travelling Salesman Problem Model with Interval-Valued Fuzzy Soft Sets with Arithmetic Mean Method

Ruba Almahasneh and László T. Kóczy

Abstract The Traveling Salesman Problem (TSP) is an extensively studied NP-hard graph search problem. Many researchers pursued the most efficient and practical solutions, by applying various techniques to find the optimum or semi optimum solution (the one with least cost). There are numerous practical extensions and modifications of the original problem, such as The Time Dependent Traveling Salesman Problem (TD TSP). Indeed, the TD TSP was towards more realistic assessment of the traffic conditions of the original TSP. The edges between nodes are assigned different costs (weights), whether they are traveled during the rush hour periods or if they crossed the traffic jam regions (such as city centers). In the classic TD TSP, the edges are assigned higher costs using concrete numbers, which might be looked at as a limitation; because those jam factors are non-deterministic and better be represented by fuzzy numbers. In this paper we introduce a more realistic novel fuzzy-based extension, the IVFSSTD TSP (Interval-Valued Fuzzy Soft Set for the Time Dependent Traveling Salesman Problem). Our core concept employs interval-valued fuzzy soft sets on the costs between nodes to realistically quantify the traffic jam regions, and the rush hours periods effects on any tour, then we user the arithmetic mean operator to take in account all factors affecting an edge simultaneously, which lead to less information loss and more adequate representation for the jam factors. Since the interval-valued fuzzy soft sets are generalization of the original fuzzy sets, which has the ability to simulate uncertain road conditions more efficiently than concrete numbers, then our approach can be considered a useful extension and a practical alternative model of the original abstract problem.

R. Almahasneh (✉) · L. T. Kóczy
Department of Telecommunications and Media Informatics, Budapest University of Technology and Economics, Budapest, Hungary
e-mail: mahasnehr@tmit.bme.hu

L. T. Kóczy
e-mail: koczy@tmit.bme.hu

L. T. Kóczy
Department of Information Technology, Szechenyi Istvan University, Gyor, Hungary

L. T. Kóczy et al. (eds.), *Computational Intelligence and Mathematics for Tackling Complex Problems 6*, Studies in Computational Intelligence 1222,
https://doi.org/10.1007/978-3-031-97879-1_5

Keywords Rush hours · Jam regions · Interval-valued fuzzy soft sets · Time Dependent Traveling Salesman Problem

1 Introduction

The Traveling Salesman Problem (TSP) originally attempts to find the shortest route starting from the company headquarter (departure point) so that all visited nodes (whether that was cities, shops or other locations) on the agenda are visited exactly once and then the salesman returns to the starting point [1]. A realistic extension of the TSP is the Time Dependent Traveling Salesman Problem (TD TSP), where the costs of the edges vary depending on whether they were crossed the jam regions or travelled during the rush hours periods [2]. In fact, the TD TSP proposed a unique concept using crisp numbers to quantify traffic jam factors which is considered a step forward towards road costs determination [2]. The TD TSP assigned fixed parts (area) of the graph as the traffic jam region in the city centers, which is far from reality. Because this is not how jam regions forms in real life; jam regions can build up gradually. They can form due to accidents, approaching the city centers or due to constructions or detours. Thus, it is more practical to represent their uncertain naturel using fuzzy numbers rather than concrete numbers. Realizing the limitation and impreciseness of such representation, we took the lead in a series of extended fuzzy-based TD TSP models. Previously, we have proposed three novel approaches. First, the Triple Fuzzy Time Dependent Traveling Salesman Problem (3FTD TSP) model [3], it is called triple fuzzy as it has three aspects of uncertainty, all represented by type-1 fuzzy sets (characterized by two-dimensional membership functions). Since the costs between the nodes are expressed by fuzzy sets, they may depend on the time and some other non-deterministic factors (such as unexpected situations as accidents, road constructions, weather conditions, etc.). This model expressed the uncertain costs added by the jam situations and rush hours and hence, calculated the overall tour length (cost) more adequately.

Second, we introduced an even more general model that allowed road uncertainties representation to become more flexible [4]. This model was called Intuitionistic Fuzzy Time Dependent Travelling Salesman Problem (IFTDT SP) [4]. This model employs type-2 intuitionistic fuzzy sets (characterized by three-dimensional membership functions) [4]. It involves the so-called hesitation part between the membership functions and the non-membership functions representing the same uncertain components in the previous model (costs, areas of traffic jams, rush hour times) allowing a higher level of vagueness representation [5–7]. The use of intuitionistic fuzzy sets in the IFTD TSP model ensured successful representation for higher degrees of association and the lower degrees of non-association for the jam factors and the rush hours and lower degrees of hesitation to edges' costs in any proposed tour, that resulted in a more accurate overall cost estimation for any given tour [4].

Third, we presented the Interval-Valued Intuitionistic Fuzzy Time Dependent Traveling Salesman Problem (IVIFTD TSP) model [8] in pursuit of considering the average collective costs of multiple factors affecting a single edge simultaneously. The IVIFTD TSP [8] uses type-2 interval-valued intuitionistic fuzzy sets [9, 10]. The improvement we achieved in this model was employing the aggregation concept of all the costs depending on their importance rather than using the max–min composition of the fuzzy factors [4] as in the IFTD TSP. This formulation led to an even more adequate model. Because in an actual tour, an edge can be affected by several factors simultaneously, and thus, by taking in consideration all the external traffic factors affecting any edge collectively at any given time, less information loss was guaranteed [8] which ultimately led to more accurate costs calculation.

In this paper, we are moving one step forwards, by proposing another practical extension that facilitates the interval-valued fuzzy soft sets to calculate the costs between nodes to realistically quantify traffic the jam regions, and the rush hours periods effects (in terms of costs) on any tour.

2 Interval-Valued Fuzzy Soft Sets

Zadeh [11] introduced type-1 fuzzy sets in 1965, their membership functions are only (n + 1)-dimensional sets. on the other hand, type-2 fuzzy sets are (n + 2)-dimensional, thus, type-2 fuzzy sets allowed more degrees of freedom in representing uncertainty while simulating any phenomenon that contains a considerable amount of vagueness or inconsistency [12, 13]. This in turn will give better calculation when applied on real-life situations as many researchers showed [14–18]. Soft set theory was introduced by Molodtsov [19] in 1999, the soft set theory has been applied in several domains, such as smoothness of functions, game theory, operations research, Riemann integration and probability. Afterwards, Maji et al. [20, 21] defined many operations for soft sets to make a detailed theoretical study on them. Also, Maji et al. successfully applied the fuzzy soft set in decision-making problems [22, 23]. Hence, soft sets theory has been studied increasingly [24] especially when the uncertainty is part of the problem's nature and formulation (such as the one in hand) [21, 22]. Yang et al. [25] defined the operations on fuzzy soft sets, which are based on three fuzzy logic operators: negation, triangular norm and triangular conorm. Zou et al. introduced the soft set and fuzzy soft set into the incomplete environment [26]. Also, Cagman et al. [27, 28] defined the concept of fuzzy parameterized fuzzy soft set (FPFS-set). Our model (the IVFSTD TSP) combines the interval-valued fuzzy set and fuzzy soft sets, to simulate the effect of jam regions and rush hours tangibly in any tour and eventually numerically measure their effects with respect to travelling costs, which fills a gap in the currently published research in this domain.

3 Interval-Valued Fuzzy Soft Time Dependent Traveling Salesman Problem (IVFSSTD TSP) Model with Arithmetic Mean

We will first introduce some definitions and operations that will be used in later sections.

3.1 *Preliminary Definitions*

We will first introduce some definitions and operations that will be used in later sections in a more analytical formal model.

3.1.1 Interval-Valued Fuzzy Sets

Let X denote a universal set $X\ (\neq \emptyset)$ with a generic element of X denoted by x, that is, $X = \{x\}$. Then the membership function by which a fuzzy set A is defined as:

$$\mu_A : X \to [0, 1] \tag{1}$$

where $[0, 1]$ denotes the interval of real numbers from 0 to 1, inclusive. It is the degree of membership of the element x to the set A. Then a fuzzy set A has the form by which μ_A is the membership function and x is a particular element of X:

$$A = \{\langle x, \mu_A(x)\rangle | x \in X\} \tag{2}$$

An interval-valued fuzzy set (IVFS) A defined on universe X is given by:

$$\tilde{A} = \left\{\left\langle x, \left[\mu_A^{\mathrm{L}}(x), \mu_A^{\mathrm{U}}(x)\right]\right\rangle | \forall x \in X\right\} \tag{3}$$

$$\tilde{\mu}_A(x) = \mu_A^{\mathrm{L}}, \mu_A^{\mathrm{U}} : X \to [0, 1] | \forall x \in X, \mu_A^{\mathrm{L}}(x) \leq \mu_A^{\mathrm{U}}(x) \tag{4}$$

where

$$\tilde{\mu}_A(x) = \left[\mu_A^{\mathrm{L}}(x), \mu_A^{\mathrm{U}}(x)\right], \ \forall x \in X$$

Let $D[0, 1]$ be the set of all closed subintervals of the interval $[0, 1]$. Therefore, for $x \in X$, $\tilde{\mu}_A(x) \in D[0, 1]$. Thus,

$$\tilde{A} = \{(x, \tilde{\mu}_A(x))\}, \widetilde{\mu}_A(x) : X \to D[0, 1], \ x \in X \tag{5}$$

Then, interval $\widetilde{\mu}_A(x) = \left[\mu_A^{\mathrm{L}}(x), \mu_A^{\mathrm{U}}(x)\right]$, is assigned for each element $x \in X$, where μ_A^{L} and μ_A^{U} are the lower extreme and the upper extreme, respectively, of the generalized membership function $\widetilde{\mu}_A$.

$$\mu_A^{\mathrm{L}}, \mu_A^{\mathrm{U}} : X \rightarrow [0, 1] | \forall x \in X, \mu_A^{\mathrm{L}}(x) \leq \mu_A^{\mathrm{U}}(x) \tag{6}$$

Thus,

$$\widetilde{\mu}_A(x) = \left[\mu_A^{\mathrm{L}}(x), \mu_A^{\mathrm{U}}(x)\right], \ \forall x \in X$$

3.1.2 Soft Sets

Soft set theory was introduced by Molodtsov [19] in 1999 as a new mathematical theory for dealing with uncertainty, it has been applied in several domains, such as smoothness of functions, game theory, operations research, Riemann integration and probability. In fact, Molodtsov applied this theory in many areas [29–31], and then formulated the notions of soft number, soft derivative, soft integral, etc. in [30–32].

On the other hand, Maji et al. [20] worked on the application of soft set in the decision making problem using the reduction of rough sets [33]. Chen et al. [23] proposed parametrization reduction of soft sets, and then Kong et al. [34] presented the normal parametrization reduction of soft sets.

This growing interest led many researchers to study the properties and applications on the soft set theory [35]. In contrast with the axiomatic approach in the classical probability theory, the soft probability is directly based on measurements [36, 37]. The soft probability may be applied to any event, in contrast to the classical probability which may be applied only to stochastically stable events [38, 39]. If the event is stochastically stable, then the soft probability will be a sufficiently narrow interval with a large trust degree if n is large enough. If the event is not stochastically stable, then the soft probability may be a large interval. The proposed approach can also be applied to construct a theory of soft clusters [40, 41].

In this section, we present the basic definitions of soft set theory that are useful for subsequent discussions. U is an initial universe, E is a set of parameters, $P(U)$ is the power set of U, and $A \subseteq E$. A soft set F_A over U is a set defined by a function f_A representing a mapping

$$f_A : E \rightarrow P(U) \text{ such that } f_A(x) = \emptyset \ if \ \ x \notin A \tag{7}$$

Here, f_A is called approximate function of the soft set F_A, and the value $f_A(x)$ is a set called *x-element* of the soft set for all $x \in E$. It is worth noting that the sets $f_A(x)$ may be arbitrary, empty, or have nonempty intersection. Thus a soft set over U can be represented by the set of ordered pairs. Also, the set of all soft sets over U will be denoted by $S(U)$.

$$F_A = \{x, f_A(x) : x \in E, f_A(x) \in P(U)\} \tag{8}$$

3.1.3 Fuzzy Soft Sets

Soft Type-2 fuzzy sets allowed more degrees of freedom in representing uncertainty while simulating any phenomenon that contains a considerable amount of vagueness or inconsistency. This in turn gave better calculation when applied on real-life situations as many researchers showed. After Molodtsov [19] introduced the Soft set, Maji et al. [21] presented the concept of the fuzzy soft sets, he also defined many operations for soft sets to make a detailed theoretical study on them. Also, Maji et al. successfully applied the fuzzy soft set in decision-making problems [21]. Mukherjee and Chakraborty [42] worked on intuitionistic fuzzy soft relations. Yang et al. [43] defined the operations on fuzzy soft sets which are based on three fuzzy logic operators: negation, triangular norm and triangular conorm. Zou and Xiao [26, 44] introduced the soft set and fuzzy soft set into the incomplete environment. Also, Cagman et al. [27, 28] defined the concept of fuzzy parameterized fuzzy soft set (FPFS-set).

Let U be a universal set, E a set of parameters and $A \subset E$. Let $F(U)$ denotes the set of all fuzzy subsets of U. Then a pair (F, A) is called a fuzzy soft set over U, where F is a mapping from A to $F(U)$.

A fuzzy soft set $\tilde{F}_A$ over U is a set defined by a function $\tilde{\mu}_A$ that represents the mapping

$$\tilde{\mu}_A(x) : E \rightarrow F(U) \tag{9}$$

such that

$$\tilde{\mu}_A(x) = \emptyset \ if \ x \notin A \tag{10}$$

Here, $\tilde{\mu}_A$ is called fuzzy approximate function of the fuzzy set $\tilde{\mathrm{F}}_A$, and the value $\tilde{\mu}_A(x)$ is a set called *x-element* of the fuzzy soft sets for all $x \in E$. Thus, $\tilde{\mathrm{F}}_A$ over U can be represented by the set of ordered pairs

$$\tilde{\mathrm{F}}_A = \{(x, \tilde{\mu}_A(x)) : x \in E, \tilde{\mu}_A(x) \in F(U)\} \tag{11}$$

all the fuzzy sets over U will be denoted by $F(U)$ and the set of all fuzzy soft-sets over U will be denoted by $FS(U)$.

3.1.4 Interval-Valued Fuzzy Soft Sets

Let U be an initial universal set, and E be a set of parameters [9, 45, 46]. A pair $(\tilde{F}, E)$ is called an interval-valued fuzzy soft set over $\tilde{P}(U)$, where $\tilde{F}$ is a mapping given by:

$$\tilde{F} : E \to \tilde{P}(U) \tag{12}$$

An interval-valued fuzzy soft set is a parameterized family of interval-valued fuzzy subsets of U, thus, its universe is the set of all interval-valued fuzzy sets of U (i.e. $\tilde{P}(U)$). An interval-valued fuzzy soft set is also a special case of a soft set because it is still a mapping from parameters to $\tilde{P}(U)$. $\forall e \in E$, $\tilde{F}(e)$ is referred as the interval fuzzy value set of parameter e, it is actually an interval-valued fuzzy set of U where $x \in U$ and $e \in E$, it can be written as:

$$\tilde{F}(e) = \left\{\left\langle x, \mu_{\tilde{F}(e)}(x)\right\rangle : x \in U : E \to \tilde{P}(U)\right\} \tag{13}$$

$\tilde{F}(e)$ is the interval-valued fuzzy membership degree that object x holds on parameter e. If $\forall e \in E$ and $\forall x \in U$, $\mu^L_{\tilde{F}(e)}(x) = \mu^U_{\tilde{F}(e)}(x)$, then $\tilde{F}(e)$ will degenerated to be a standard fuzzy set and then $(\tilde{F}, E)$ will be degenerated to be a traditional fuzzy soft set.

4 The Basic Model Methodology

To apply this methodology on a simple tour for the TD TSP under rush hours and jam regions. Let us assume we have 12 nodes (Fig. 1).

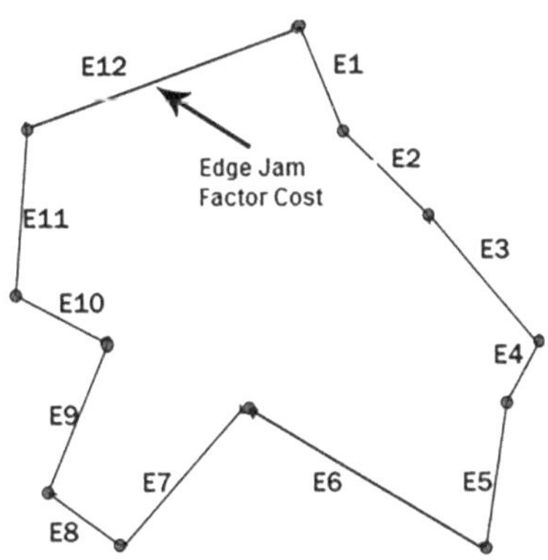

Fig. 1 Tour for a simple example

Suppose J is the set of the jam factors (universe). C denotes the set of the modified costs (parameters), and E denotes the set of the edges in a tour. Construct an interval-valued fuzzy soft set (F, C) over J. $\tilde{\mathrm{F}}(J)$ is the set of all interval valued fuzzy sets of J the mapping given by:

$$F : C \rightarrow \tilde{\mathrm{F}}(J) \tag{14}$$

A relation matrix R_1 is constructed from the interval-valued fuzzy soft set (F, C) and called jam factors-costs matrix. Similarly, its complement $(F, C)^c$ gives another relation matrix, R_2, called non-jam factors-costs matrix.

We construct another interval-valued fuzzy soft set (F_1, J) over E, where $\tilde{\mathrm{F}}(E)$ is a set of all interval valued fuzzy sets of E, the mapping given by:

$$F_1 : J \rightarrow \tilde{\mathrm{F}}(E) \tag{15}$$

This interval-valued fuzzy soft set gives another relation matrix Q called edges-jam factors matrix. Here we apply Sanchez's method for the arithmetic mean of interval valued fuzzy matrix.

Let $A = [AL, AU]$. Arithmetic mean of an IVFM A is the arithmetic mean of AL and AU denoted as $\mathrm{am}(A)$ is defined as $\mathrm{am}(A) = \frac{AL+AU}{2}$. By using the arithmetic mean of an IVFM, we compute the $\mathrm{am}(R1)$, $\mathrm{am}(R2)$ and $\mathrm{am}(Q)$ for the matrices:

$$R1 = [R1L, R1U],\ \ R2 = [R2L, R2U]\ and\ Q = \left[QL, QU\right].$$

Thus,

$$\mathrm{am}(R1) = \frac{R1L + R1U}{2} \tag{16}$$

$$\mathrm{am}(R2) = \frac{R2L + R2U}{2} \tag{17}$$

$$\mathrm{am}(Q) = \frac{QL + QU}{2} \tag{18}$$

Then combining the relation matrices $\mathrm{am}(R1)$ and $\mathrm{am}(R2)$ separately with $\mathrm{am}(Q)$ under the MAX.MIN composition of fuzzy matrices we get

$$T1 = \mathrm{am}(Q) \circ \mathrm{am}(R1)$$

$$T2 = \mathrm{am}(Q) \circ \mathrm{am}(R2)$$

$$T3 = \mathrm{am}(Q) \circ \mathrm{am}(R1)^c$$

$$T4 = \mathrm{am}(Q) \circ \mathrm{am}(R2)^c$$

$$\mathrm{am}(R_1) = \begin{bmatrix} [0.85\ 0.75] \\ [0.25\ 0.5] \\ [0.55\ 0.45] \\ [0.3\ 0.45] \end{bmatrix} \quad \mathrm{am}(R_2) = \begin{bmatrix} [0.15\ 0.25] \\ [0.75\ 0.5] \\ [0.6\ 0.6] \\ [0.7\ 0.1] \end{bmatrix}$$

$$\mathrm{am}(Q) = \begin{bmatrix} [0.75\ 0.4\ 0.9\ 0.75] \\ [0.4\ 0.5\ 0.3\ 0.4] \\ [0.7\ 0.4\ 0.6\ 0.3] \\ [0.75\ 0.4\ 0.9\ 0.75] \\ [0.4\ 0.5\ 0.3\ 0.4] \\ [0.4\ 0.5\ 0.3\ 0.4] \\ [0.75\ 0.4\ 0.9\ 0.75] \\ [0.4\ 0.5\ 0.3\ 0.4] \\ [0.4\ 0.5\ 0.3\ 0.4] \\ [0.75\ 0.4\ 0.9\ 0.75] \\ [0.4\ 0.5\ 0.3\ 0.4] \\ [0.4\ 0.5\ 0.3\ 0.4] \end{bmatrix}$$

Then combining the relation matrices am($R1$) and am($R2$) separately with am(Q) we have:

$$T_1 = \mathrm{am}(Q) \circ \mathrm{am}(R1) = \begin{pmatrix} [0.75\ 0.75] \\ [0.4\ 0.5] \\ [0.7\ 0.7] \\ [0.75\ 0.75] \\ [0.4\ 0.5] \\ [0.7\ 0.7] \\ [0.75\ 0.75] \\ [0.4\ 0.5] \\ [0.7\ 0.7] \\ [0.75\ 0.75] \\ [0.4\ 0.5] \\ [0.7\ 0.7] \end{pmatrix}$$

$$T_2 = \mathrm{am}(Q) \circ \mathrm{am}(R2) = \begin{pmatrix} [0.7\ 0.55] \\ [0.5\ 0.5] \\ [0.7\ 0.55] \\ [0.7\ 0.55] \\ [0.5\ 0.5] \\ [0.7\ 0.55] \\ [0.7\ 0.55] \\ [0.5\ 0.5] \\ [0.7\ 0.55] \\ [0.7\ 0.55] \\ [0.5\ 0.5] \\ [0.7\ 0.55] \end{pmatrix}$$

$$T_3 = \mathrm{am}(Q) \circ \mathrm{am}(R1)^c = \begin{pmatrix} [0.75\ 0.55] \\ [0.5\ 0.5] \\ [0.45\ 0.55] \\ [0.75\ 0.55] \\ [0.5\ 0.5] \\ [0.45\ 0.55] \\ [0.75\ 0.55] \\ [0.5\ 0.5] \\ [0.45\ 0.55] \\ [0.75\ 0.55] \\ [0.5\ 0.5] \\ [0.45\ 0.55] \end{pmatrix}$$

$$T_4 = \mathrm{am}(Q) \circ \mathrm{am}(R2)^c = \begin{pmatrix} [0.75\ 0.75] \\ [0.4\ 0.5] \\ [0.7\ 0.7] \\ [0.75\ 0.75] \\ [0.4\ 0.5] \\ [0.7\ 0.7] \\ [0.75\ 0.75] \\ [0.4\ 0.5] \\ [0.7\ 0.7] \\ [0.75\ 0.75] \\ [0.4\ 0.5] \\ [0.7\ 0.7] \end{pmatrix}$$

we calculate the edges score $\boldsymbol{S}_{\boldsymbol{T}_1}$ and $\boldsymbol{S}_{\boldsymbol{T}_2}$ for and against the costs respectively as follows:

Table 1 The modified cost for edges

$S_{T_1} - S_{T_2}$	c_1	c_2
e_1	0.05	0.15
e_2	– 0.1	0
e_3	0.25	0.15
e_4	0.05	0.15
e_5	– 0.1	0
e_6	0.25	0.15
e_7	0.05	0.15
e_8	– 0.1	0
e_9	0.25	0.15
e_{10}	0.05	0.15
e_{11}	– 0.1	0
e_{12}	0.25	0.15

$S_{T1} = \max_{i,j}\left(\left[T_1\left(e_i, c_j\right), T_4\left(e_i, c_j\right)\right]\right)$, in our example, $(i = 1, 2, \ldots 12)$ the edges count and $(j = 1, 2)$ for the costs.
$S_{T2} = \max_{i,j}\left(\left[T_2\left(e_i, c_j\right), T_3\left(e_i, c_j\right)\right]\right)$, in our example the edges count $(i = 1, 2, \ldots 12)$ and $(j = 1, 2)$ for the costs.

$$S_{T1} = \begin{pmatrix} [0.75\ 0.75] \\ [0.4\ 0.5] \\ [0.7\ 0.7] \\ [0.75\ 0.75] \\ [0.4\ 0.5] \\ [0.7\ 0.7] \\ [0.75\ 0.75] \\ [0.4\ 0.5] \\ [0.7\ 0.7] \\ [0.75\ 0.75] \\ [0.4\ 0.5] \\ [0.7\ 0.7] \end{pmatrix} \qquad S_{T2} = \begin{pmatrix} [0.75\ 0.55] \\ [0.5\ 0.5] \\ [0.45\ 0.55] \\ [0.75\ 0.55] \\ [0.5\ 0.5] \\ [0.45\ 0.55] \\ [0.75\ 0.55] \\ [0.5\ 0.5] \\ [0.45\ 0.55] \\ [0.75\ 0.55] \\ [0.5\ 0.5] \\ [0.45\ 0.55] \end{pmatrix}$$

In Table 1, we calculate the modified cost for each edge we apply $[S_{T1} - S_{T2}]$.

5 Conclusions

We have applied the notion of interval-valued fuzzy soft sets with the arithmetic mean operator to cater for all jam factors that are effecting any edge in a tour simultaneously, this adequately calculated the overlapping jam factors for an edge to calculate the overall added cost tangibility with least information loss. The case study has been taken to exhibit the simplicity of the technique and applicability. In our future work we are eager to test and validate this approach on the largest instances in the literature and compare the results with our previously extended fuzzy approaches for the TD TSP with the jam factors and the rush hours and with the state of the art solutions.

References

1. Applegate, D.L., Bixby, R.E., Chvátal, V., Cook, W.J.: The Traveling Salesman Problem: A Computational Study, pp. 1–81. Princeton University Press, Princeton
2. Schneider, J.: The time-dependent traveling salesman problem. Phys. A **314**, 151–155 (2002)
3. Kóczy, L.T., Földesi, P., Tüű-Szabó, B., Almahasneh, R.: Modeling of fuzzy rule-base algorithm for the time dependent traveling salesman problem. In: Proceedings of the IEEE International Conference on Fuzzy Systems (FUZZ-IEEE) (2019)
4. Almahasneh, R., Tüű-Szabó, B., Földesi, P., Kóczy: Intuitionistic fuzzy model of jam regions and rush hours for the time dependent traveling salesman problem. In: Proceedings of the IFSA World Congress and NAFIPS Annual Conference (2019)
5. Atanassov, R.K.: Intuitionistic fuzzy sets. Fuzzy Sets Syst. **20**, 87–96 (1986)
6. Biswas, R.: On fuzzy sets and intuitionistic fuzzy sets. Notes Intuition. Fuzzy Sets **3**, 3–11 (1997)
7. Gorzalczany, M.B.: Approximate inference with interval-valued fuzzy sets—an outline. In: Proceedings of the Polish Symposium on Interval and Fuzzy Mathematics, Poznafi, Poland, pp. 89–95 (1983)
8. Almahasneh, R., Tüű-Szabó, B., Földesi, P., Kóczy: Interval-valued intuitionistic fuzzy model of jam regions and rush hours for the time (2020)
9. Yang, X., Lin, T.Y., Yang, J., Li, Y., Yu, D.: Combination of interval-valued fuzzy set and soft set. Comput. Math. Appl. **58**, 521–527 (2009)
10. Short remark on fuzzy sets, interval type-2 fuzzy sets, general type-2 fuzzy sets and intuitionistic fuzzy sets. In: IEEE Conference on Intelligent Systems, vol. 1, pp. 183–190 (2014)
11. Zadeh, L.A.: Fuzzy sets. Inf. Control **8**(3), 338–353 (1965)
12. Mamdani, E.H.: Application of fuzzy algorithms for control of simple dynamic plant. IEEE Proc. **121**(12), 1585–1588 (1974)
13. Hirota, K.: Concept of probabilistic sets. Fuzzy Sets Syst. **5**, 31–46 (1981)
14. Mendel, J.M., Bob John, R.I., Liu, F.: Interval type-2 fuzzy logic systems made simple. IEEE Trans. Fuzzy Syst. **14**(6) (2006)
15. Mendel, J.M., Bob John, R.I.: Type-2 fuzzy sets made simple. IEEE Trans. Fuzzy Syst. **10**(2) (2002)
16. Czogala, E.: On distribution function description of probabilistic sets and its application in decision making. Fuzzy Sets Syst. **10**, 21–29 (1983)
17. Mittal, K., Jain, A., Vaisla, K.S., Castillo, O., Kacprzyk, J.: A comprehensive review on type 2 fuzzy logic applications: past, present and future. Eng. Appl. Artif. Intell. **95** (2020)
18. Karmakar, S., Seikh, M.R., Castillo, O.: Type-2 intuitionistic fuzzy matrix games based on a new distance measure: application to biogas-plant implementation problem. Appl. Soft Comput. **106** (2021)
19. Molodtsov, D.A.: Soft set theory—first results. Comput. Math. Appl. **37**, 19–31 (1999)
20. Maji, P.K., Biswas, R., Roy, A.R.: Soft set theory. Comput. Math. Appl. **45**, 555–562 (2003)
21. Maji, P.K., Biswas, R., Roy, A.R.: Fuzzy soft sets. J. Fuzzy Math. **9**, 589–602 (2001)
22. Maji, P.K., Roy, A.R., Biswas, R.: An application of soft sets in a decision making problem. Comput. Math. Appl. **44**, 1077–1083 (2002)

23. Chen, D., Tsang, E.C.C., Yeung, D.S., Wang, X.: The parameterization reduction of soft sets and its applications. Comput. Math. Appl. **49**, 757–763 (2005)
24. Maji, P.K., Biswas, R., Roy, A.R.: Fuzzy soft sets. J. Fuzzy Math. **9**(3), 589–602 (2001)
25. Yang, X.B., Yu, D.J., Yang, J.Y., Wu, C.: Generalization of soft set theory: from crisp to fuzzy case. In: Cao, B.Y. (ed.) Proceeding of the Second International Conference on Fuzzy Information and Engineering, Guangzhou, 16–23 May 2007. Advance on Soft Computing, vol. 40, pp. 345–354. Springer-Verlag, Berlin, Heidelberg (2007)
26. Zou, Y., Xiao, Z.: Data analysis approaches of soft sets under incomplete information. Knowl.-Based Syst. **21**, 941–994 (2008)
27. Cagman, N., Enginoglu, S.: Soft matrices and its decision makings. Comput. Math. Appl. **59**, 3308–3314 (2010)
28. Cagman, N., Citak, F., Enginoglu, S.: Fuzzy parameterized fuzzy soft set theory and its applications. Turk. J. Fuzzy Syst. **1**, 21–35 (2010)
29. Molodtsov, D.A.: Soft set theory-first results. Comput. Math. Appl. **37**, 19–31 (1999)
30. Molodtsov, D.A.: The description of a dependence with the help of soft sets. J. Comput. Syst. Sci. Int. **40**(6), 977–984 (2001)
31. Molodtsov, D.A.: The Theory of Soft Sets. URSS Publishers, Moscow (2004). (in Russian)
32. Molodtsov, D.A., Leonov, V.Yu., Kovkov, D.V.: Soft sets technique and its application. Nechetkie Sist. Myakie Vychisleniya **1**(1), 8–39 (2006)
33. Pawlak, Z.: Rough sets. Int. J. Comput. Inf. Sci. **11**, 341–356 (1982)
34. Kong, Z., Gao, L., Wang, L.: Comment on "A fuzzy soft set theoretic approach to decision making problems." J. Comput. Appl. Math. **223**, 540–542 (2009)
35. Xiao, Z., Li, Y., Zhong, B., Yang, X.: Research on synthetically evaluating method for business competitive capacity based on soft set. Stat. Methods Med. Res. 52–54 (2003)
36. Xiao, Z., Chen, L., Zhong, B., Ye, S.: Recognition for soft information based on the theory of soft sets. In: Chen, J. (eds.) Proceedings of ICSSSM-05, vol. 2, pp. 1104–1106 (2005)
37. Roy, A.R., Maji, P.K.: A fuzzy soft set theoretic approach to decision making problems. J. Comput. Appl. Math. **203**, 412–418 (2007)
38. Mushrif, M.M., Sengupta, S., Ray, A.K.: Texture classification using a novel, soft-set theory based classification, algorithm. Lect. Notes Comput. Sci. **3851**, 246–254 (2006)
39. Pei, D., Miao, D.: From soft sets to information systems. In: Hu, X., Liu, Q., Skowron, A., Lin, T.Y., Yager, R.R., Zhang, B. (eds.) Proceedings of Granular Computing, vol. 2, pp. 617–621. IEEE (2005)
40. Som, T.: On the theory of soft sets, soft relation and fuzzy soft relation. In: Proceedings of the National Conference on Uncertainty: A Mathematical Approach, UAMA-06, Burdwan, pp. 1–9 (2006)
41. Sun, Q.M., Zhang, Z.L., Liu, J.: Soft sets and soft modules. In: Wang, G., Li, T., Grzymala-Busse, J.W., Miao, D., Skowron, A., Yao, Y. (eds.) Rough Sets and Knowledge Technology, RSKT-08, Proceedings, pp. 403–409. Springer (2008)
42. Mukherjee, A., Chakraborty, S.B.: On intuitionistic fuzzy soft relations. Bull. Kerala Math. Assoc. **5**(1), 35–42 (2008)
43. Yang, X., Yu, D., Yang, J., Wu, C.: Generalization of soft set theory: from crisp to fuzzy case. In: Cao, B.-Y. (ed.) Fuzzy Information and Engineering: Proceedings of ICFIE2007. Advances in Soft Computing, vol. 40, pp. 345–355. Springer (2007)
44. Xiao, Z., Gong, K., Zou, Y.: A combined forecasting approach based on fuzzy soft sets. J. Comput. Appl. Math. **228**, 326–333 (2009)
45. Kong, Z., Gao, L., Wang, L., Li, S.: The normal parameter reduction of soft sets and its algorithm. Comput. Math. Appl. **56**, 3029–3037 (2008)
46. Zou, Y., Chen, Y.: Research on soft set theory and parameters reduction based on relational algebra. In: Intelligent Information Technology Application, IITA, vol. 1, pp. 152–156 (2008)

On Aggregation Operators and Utility Functions with Truncation Property

Dragan Jočić and Ivana Štajner-Papuga

Abstract Aggregation operators form a vast class of mathematical tools that are highly applicable in modelling decision-making processes. Of special interest are aggregation operators with an annihilator, i.e., with an absorbing element, since they can be applicable in modelling situations involving participants with the power of veto. This paper presents an overview of results that consider the conditional distributivity of such operators, namely T-uninorms and nullnorms, over t-conorms and an analysis of the role of obtained distributive pairs in constructions of utility functions with a truncation property.

Keywords Aggregation operators · Annihilator · T-uninorms · Nullnorms · Conditional distributivity · Utility function

1 Introduction

Aggregation operators represent a general mathematical notion that is recognized in many research areas due to their applicability in modelling both theoretical and practical problems, from natural sciences and mathematics to economics and social sciences. The basic definition, i.e., the skeleton of aggregation operators consists of monotonicity and boundary conditions, however, depending on the specificity of the observed problem, some additional properties can be, and usually are, requested. Of special interest for researchers are pairs of distributive operators, and some of those additional properties derive from the solutions of distributivity equations. As it is

This research was supported by the Science Fund of the Republic of Serbia, Grant No. 6565, Advanced Techniques of Mathematical Aggregation and Approximative Equations Solving in Digital Operational Research-AT-MATADOR.

D. Jočić
Mathematical Institute of the Serbian Academy of Sciences and Arts, Belgrade, Serbia

I. Štajner-Papuga (✉)
Department of Mathematics and Informatics, Faculty of Sciences, University of Novi Sad, Novi Sad, Serbia
e-mail: ivana.stajner-papuga@dmi.uns.ac.rs

L. T. Kóczy et al. (eds.), *Computational Intelligence and Mathematics for Tackling Complex Problems 6*, Studies in Computational Intelligence 1222,
https://doi.org/10.1007/978-3-031-97879-1_6

well known, investigation of distributivity has roots in [1], and on the full domain, the inner aggregation operator reduces to an idempotent operator (see [4, 6, 7]). Interestingly enough, this problem can be avoided by restricting the domain of observed aggregation operators. By excluding some border cases, a relaxed version of distributivity was introduced under the name conditional or restricted distributivity, and with this approach, a larger variety of solutions can be obtained (see [3, 4, 6, 7]). Additionally, the concept of conditional distributivity has proven to be very useful in utility theory (see [2, 5]).

This paper provides an overview of conditionally distributive pairs of aggregation operators with annihilator and continuous t-conorm, as well as a discussion of the application of the obtained results in utility theory. Some preliminary notions concerning aggregation operators are conditional distributivity are given in Sect. 2. An overview of results from [4–6] for conditional distributivity of T-uninorms and nullnorms over t-conorms is given in this section. An analysis of the role of conditionally distributive pairs of aggregation operators in the utility theory [2, 5] is the focus of Sect. 3. Also, properties of the utility function U_F, where F is a T-uninorm, are given in this section. Some concluding remarks are given in Sect. 4.

2 Preliminary Notions

2.1 Aggregation Operators

Definition 1 ([3]) An aggregation operator on $[0, 1]^n$ is a function $A^{(n)} : [0, 1]^n \rightarrow [0, 1]$ that is nondecreasing in each variable and that fulfills the boundary conditions

$$A^{(n)}(0, \ldots, 0) = 0 \quad \text{and} \quad A^{(n)}(1, \ldots, 1) = 1.$$

The focus of this paper is on binary aggregation operators, and the notation A will be used instead of $A^{(2)}$. Some additional properties such as continuity, associativity, commutativity, etc., can be required, depending on a type of a problem that is being modelled with aggregation operators in question [3]. Often, the existence of an annihilator (absorbing element), i.e., of an element $k \in [0, 1]$ such that $A(k, x) = A(x, k) = k$ for all $x \in [0, 1]$, is requested. At the base of the research presented here are associative commutative aggregation operators with an annihilator k (associative k-CAOA) that were studied in [8]. Further on, for any binary aggregation operator A, and some $c \in [0, 1]$, the section $A_c : [0, 1] \rightarrow [0, 1]$ of the form $A_c(x) = A(c, x)$ will be denoted by A_c. The continuity (discontinuity) of sections A_0 and A_1 has a key role in the classification of the associative k-CAOA.

Definition 2 ([8]) Let $A : [0, 1]^2 \to [0, 1]$ be an associative k-CAOA.

- A is called S-uninorm if:
  - A_0 is continuous, and A_1 is not;
  - there is an IFC element, i.e., there is an idempotent element $e \in (0, 1)$ for A, such that A_e is continuous and $A_e(1) = 1$.
- A is called T-uninorm if:
  - A_1 is continuous, and A_0 is not;
  - there is an IFC element, i.e., there is an idempotent element $e \in (0, 1)$ for A, A_e is continuous and $A_e(0) = 0$.
- A is called bi-uninorm if:
  - A_0 and A_1 are not continuous;
  - there are IFC elements, i.e., there are idempotent elements $e_0, e_1 \in (0, 1)$ for A, such that A_{e_0} and A_{e_1} are continuous, $A_{e_0}(0) = 0$, and $A_{e_1}(1) = 1$.
- A is called nullnorm (t-operator) if:
  - A_0 and A_1 are continuous.

Since the main interest of this paper are T-uninorms and nullnorms, an overview of the results from [8] for those two classes of associative k-CAOA follows. Further on let $I_k = [0, k] \times [k, 1] \cup [k, 1] \times [0, k]$.

- A binary aggregation operator $A : [0, 1]^2 \to [0, 1]$ is a T-uninorm if and only if there exists $k \in (0, 1]$, a t-norm T, and a disjunctive uninorm U' with the neutral element $e' \in (0, 1)$ such that

$$A(x, y) = \begin{cases} kU'\left(\frac{x}{k}, \frac{y}{k}\right) & \text{if } (x, y) \in [0, k]^2, \\ k + (1-k)T\left(\frac{x-k}{1-k}, \frac{y-k}{1-k}\right) & \text{if } (x, y) \in [k, 1]^2, \\ k & \text{if } (x, y) \in I_k. \end{cases} \tag{1}$$

  If e is the IFC element of a T-uninorm, then $e' = \frac{e}{k}$. The IFC element e is the neutral element for a T-uninorm on $[0, k]^2$. If the uninorm U' is from the class $U_{\max}$ (see [3]), then a T-uninorm A is called a T-uninorm in $U_{\max}$, and this is the class of T-uninorms that is covered in this paper.
- A binary aggregation operator $A : [0, 1]^2 \to [0, 1]$ is a nullnorm with an annihilator $k \in (0, 1)$ if and only if there exists a t-norm T and a t-conorm S such that A is given by

$$A(x, y) = \begin{cases} kS\left(\frac{x}{k}, \frac{y}{k}\right) & \text{if } (x, y) \in [0, k]^2, \\ k + (1-k)T\left(\frac{x-k}{1-k}, \frac{y-k}{1-k}\right) & \text{if } (x, y) \in [k, 1]^2, \\ k & \text{if } (x, y) \in I_k. \end{cases} \tag{2}$$

For $k = 0$, the nullnorm is a t-norm, while for $k = 1$ it is a t-conorm.

2.2 Conditional Distributivity

Functional equations that are called the left and right distributivity laws were introduced in [1]. As seen in [7], a t-norm is distributive over a t-conorm only for idempotent t-conorm, i.e., for t-conorm being $S_M = \max$. Also, it was shown that with exclusion of some border points from the domain of the observed aggregation operators, the number of solutions has increased. This relaxed version of distributivity concept is known as the conditionally distributive [7] or the restricted distributivity (see [3]) and its more general definition follows.

Definition 3 ([4, 6]) Let F be a T-uninorm or a nullnorm with annihilator $k \in (0, 1)$, and S be a t-conorm. F is conditionally distributive (CD) over S if for all $x, y, z \in [0, 1]$ the following holds

$$\text{(CD)} \qquad F(x, S(y, z)) = S(F(x, y), F(x, z)), \text{ whenever } S(y, z) < 1.$$

In the case when operator F is a t-norm, Definition 3 coincides with the original definition from [7].

Now, distributivity pairs are characterized by the following theorem.

Theorem 1 *([4, 6]) Let F be a continuous nullnorm or a T-uninorm in $U_{\max}$ with continuous underlying t-norm T', and S a continuous t-conorm. F is conditionally distributive over S if and only if exactly one of the following cases is fulfilled:*

1. $S = S_M = \max$;
2. *there is an $a \in [k, 1)$ such that S is of the form*

$$S(x, y) = \begin{cases} a + (1-a)S^*\left(\frac{x-a}{1-a}, \frac{y-a}{1-a}\right) & \textit{if } (x, y) \in [a, 1]^2, \\ \max(x, y) & \textit{otherwise}, \end{cases} \tag{3}$$

and F is given by

$$F(x, y) = \begin{cases} kA\left(\frac{x}{k}, \frac{y}{k}\right) & \textit{if } (x, y) \in [0, k]^2, \\ k + (a-k)T_1\left(\frac{x-k}{a-k}, \frac{y-k}{a-k}\right) & \textit{if } (x, y) \in [k, a]^2, \\ a + (1-a)T_*\left(\frac{x-a}{1-a}, \frac{y-a}{1-a}\right) & \textit{if } (x, y) \in [a, 1]^2, \\ k & \textit{if } (x, y) \in I_k, \\ \min(x, y) & \textit{otherwise}, \end{cases} \tag{4}$$

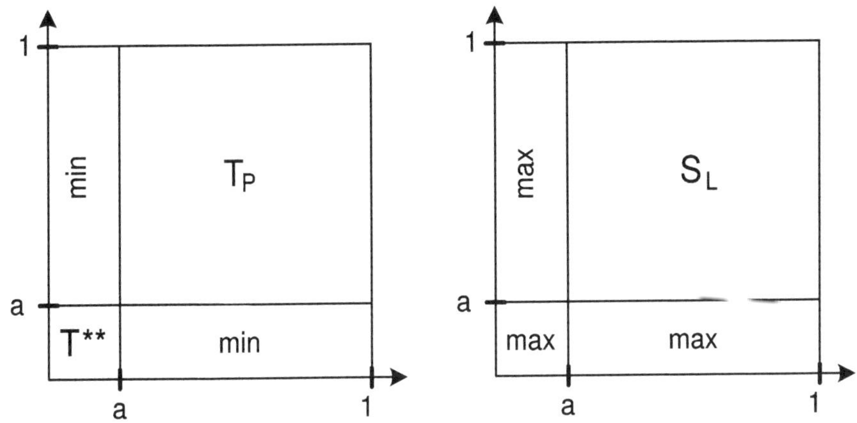

Fig. 1 Conditional distributivity: t-norm and t-conorm

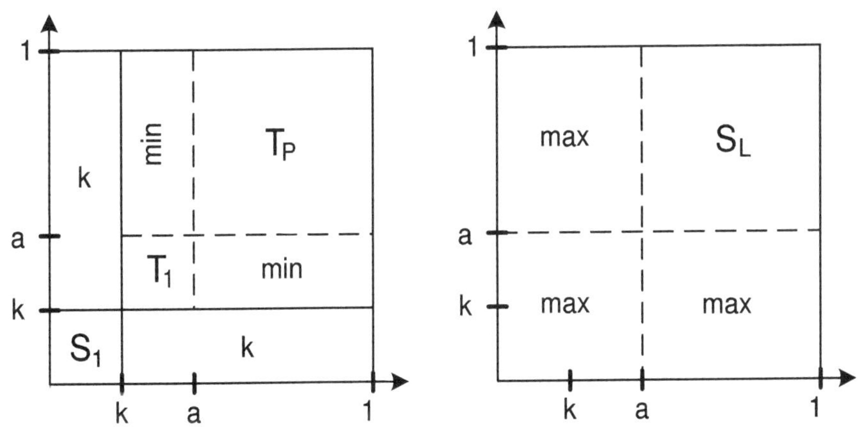

Fig. 2 Conditional distributivity: nullnorm and t-conorm

where S^* is a nilpotent t-conorm, T_* is a strict t-norm, T_1 is a continuous t-norm, and A is a uninorm from the class $U_{\max}$ when F is a T-uninorm, or an arbitrary continuous t-conorm when F is a nullnorm.

Remark 1 Due to isomorphisms between strict t-norms and product t-norm $T_P(x, y) = x \cdot y$ and nilpotent t-conorms and the Lukasiewicz t-conorm $S_L(x, y) = \min\{x + y, 1\}$ (see [7]), the previous result is often used in the form where $S_* = S_L$ and $T_* = T_P$ (see Fig. 2).

If operator F is a continuous t-norm, i.e., $k = 0$, Theorem 1 coincides with the corresponding result from [7] (see Figs. 1 and 2).

3 Utility Function

The importance of conditionally distributive pairs of aggregation operators for Utility Theory can be illustrated by the results from papers [2, 5].

Paper [2] contains an example of an optimistic hybrid utility function U_h that is constructed from a continuous t-norm T and a continuous t-conorm S that have to fulfill conditional distributivity law.

$$U_h(u_1, u_2; \mu_1, \mu_2) = S(T(u_1, \mu_1), T(u_2, \mu_2))$$

where u_1, u_2 are utilities in $[0, 1]$ and μ_1, μ_2 are degrees of plausibility from $\Phi_{S,a}$.

$$\Phi_{S,a} = \{(\alpha, \beta) | (\alpha, \beta) \in (a, 1)^2, \alpha + \beta = 1 + a, \text{ or } \\ \min(\alpha, \beta) \leq a, \max(\alpha, \beta) = 1\}.$$

This approach has been extended in [5] with an example of a utility function obtained from a continuous nullnorm F with an annihilator $k \in (0, 1)$, and a continuous t-conorm S that also satisfies the conditional distributivity law. In this case U_F is of the following form

$$U_F(u_1, u_2; \mu_1, \mu_2) = S(F(u_1, \mu_1), F(u_2, \mu_2)) \tag{5}$$

where u_1, u_2 are utilities in $[0, 1]$ and μ_1, μ_2 are degrees of plausibility from $\Phi_{S,a}$. It turns out that its value cannot be less then k, i.e., a threshold that is imposed by the absorbing element of a continuous nullnorm is present. Thus, annihilator of F allows a Decision Maker to predetermine a threshold for a certain decision process. Additionally, the behavioral characteristics of the obtained utility function are investigated in [5].

Now, taking into account Theorem 1, the question is what are properties of the utility function U_F when F is a T-uninorm. Examination of the behavioral characteristics of this new utility function follows.

Case I Let $\mu_1 > a, \mu_2 > a$, i.e., $\mu_1 + \mu_2 = 1 + a$. The following subcases have to be considered.

1. Let $u_1 > a, \ u_2 > a$. Then

$$\begin{aligned} U_F(u_1, u_2; \mu_1, \mu_2) &= S\left(a + \frac{(u_1 - a)(\mu_1 - a)}{1 - a}, a + \frac{(u_2 - a)(\mu_2 - a)}{1 - a}\right) \\ &= a + \frac{(u_1 - a)(\mu_1 - a)}{1 - a} + \frac{(u_2 - a)(\mu_2 - a)}{1 - a} \\ &= \frac{u_1(\mu_1 - a) + u_2(1 - \mu_1)}{1 - a}. \end{aligned}$$

2. Let $u_2 > a \geq u_1$. Then

$$U_F(u_1, u_2; \mu_1, \mu_2) = a + \frac{(u_2 - a)(\mu_2 - a)}{1 - a}.$$

3. Let $u_1 > a \geq u_2$. As in the previous case we have

$$U_F(u_1, u_2; \mu_1, \mu_2) = a + \frac{(u_1 - a)(\mu_1 - a)}{1 - a}.$$

4. Let $u_1 \leq a,\ u_2 \leq a$. Then:
   (a) if $u_1 \leq k,\ u_2 \leq k$, then $U_F(u_1, u_2; \mu_1, \mu_2) = S(k, k) = \max(k, k) = k$;
   (b) if $u_2 \leq k < u_1$, then $U_F(u_1, u_2; \mu_1, \mu_2) = S(u_1, k) = \max(u_1, k) = u_1$;
   (c) if $u_1 \leq k < u_2$, then $U_F(u_1, u_2; \mu_1, \mu_2) = S(k, u_2) = \max(u_2, k) = u_2$;
   (d) if $k < u_1,\ k < u_2$, then $U_F(u_1, u_2; \mu_1, \mu_2) = S(u_1, u_2) = \max(u_1, u_2)$.

In this case, the conclusion is the same as in [5], i.e., in the worst case (4a), when both rewards are lower than k, the decision maker looks forward to the reword k. Therefore, the threshold appears again. Also, in the first three subcases we have the same situation as for the hybrid utility function U from [2]. Therefore, in the subcase 1 the decision maker is probabilistic, whereas in subcases 2 and 3 the decision maker looks forward to the best outcome (Fig. 3).

Case II Let $\mu_1 \leq a, \mu_2 = 1$ (the corresponding analysis of the case $\mu_2 \leq a, \mu_1 = 1$ can be done in an analogous way). Now the following subcases are being considered.

1. Let $\mu_1 \leq k$. Then the following holds.
   (a) If $u_1 > a,\ u_2 > a$, then $U_F(u_1, u_2; \mu_1, \mu_2) = S(k, u_2) = \max(k, u_2) = u_2$.
   (b) If $u_1 \leq a,\ u_2 \leq a$, then:
      i. if $u_1 \leq k,\ u_2 \leq k$, then $U_F(u_1, u_2; \mu_1, \mu_2) = \max(U'(u_1, \mu_1), k) = k$;
      ii. if $u_2 \leq k < u_1$, then $U_F(u_1, u_2; \mu_1, \mu_2) = \max(k, k) = k$;
      iii. if $u_1 \leq k < u_2$, then $U_F(u_1, u_2; \mu_1, \mu_2) = \max(U'(u_1, \mu_1), u_2) = u_2$;
      iv. if $k < u_1,\ k < u_2$, then $U_F(u_1, u_2; \mu_1, \mu_2) = \max(k, u_2) = u_2$.

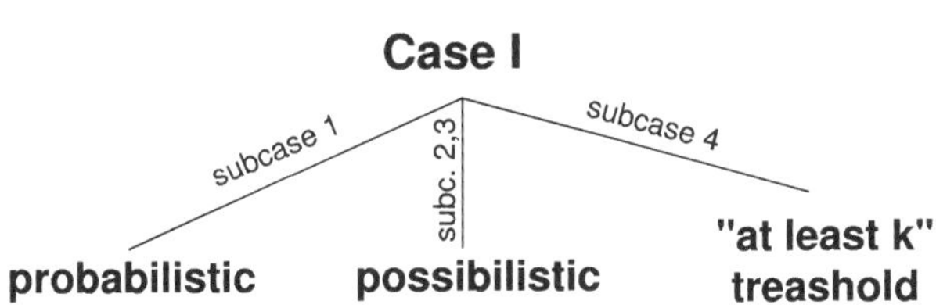

Fig. 3 Behavioral characteristics for Case I

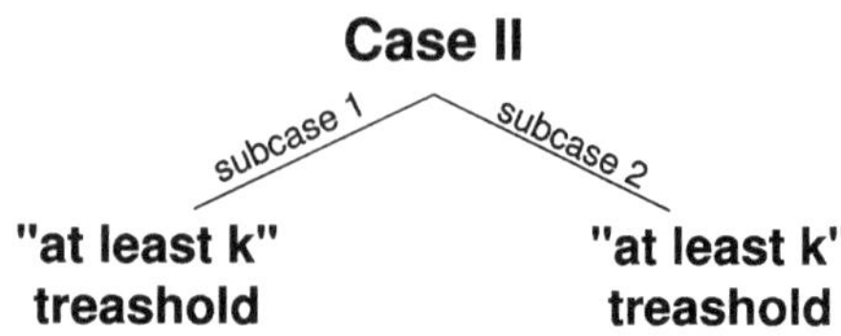

Fig. 4 Behavioral characteristics for Case II

(c) If $u_1 \leq a < u_2$, then:
   i. if $u_1 \leq k$, then $U_F(u_1, u_2; \mu_1, \mu_2) = \max(U'(u_1, \mu_1), u_2) = u_2$;
   ii. if $u_1 > k$, then $U_F(u_1, u_2; \mu_1, \mu_2) = \max(k, u_2) = u_2$.

(d) If $u_2 \leq a < u_1$, then:
   i. if $u_2 \leq k$, then $U_F(u_1, u_2; \mu_1, \mu_2) = \max(k, k) = k$;
   ii. if $u_2 > k$, then $U_F(u_1, u_2; \mu_1, \mu_2) = \max(k, u_2) = u_2$.

2. Let $\mu_1 > k$. Then the following holds.

(a) If $u_1 > a,\ u_2 > a$, then $U_F(u_1, u_2; \mu_1, \mu_2) = \max(\mu_1, u_2) = u_2$.

(b) If $u_1 \leq a,\ u_2 \leq a$, then:
   i. if $u_1 \leq k,\ u_2 \leq k$, then $U_F(u_1, u_2; \mu_1, \mu_2) = \max(k, k) = k$;
   ii. if $u_1 \leq k < u_2$, then $U_F(u_1, u_2; \mu_1, \mu_2) = \max(k, u_2) = u_2$;
   iii. if $u_2 \leq k < u_1$, then $U_F(u_1, u_2; \mu_1, \mu_2) = \max(T_1(u_1, \mu_1), k) = T_1(u_1, \mu_1)$;
   iv. if $k < u_1,\ k < u_2$, then $U_F(u_1, u_2; \mu_1, \mu_2) = \max(T_1(u_1, \mu_1), u_2)$.

(c) If $u_1 \leq a < u_2$, then:
   i. if $u_1 \leq k$, then $U_F(u_1, u_2; \mu_1, \mu_2) = \max(k, u_2) = u_2$;
   ii. if $u_1 > k$, then $U_F(u_1, u_2; \mu_1, \mu_2) = \max(T_1(u_1, \mu_1), u_2) = u_2$.

(d) If $u_2 \leq a < u_1$, then:
   i. if $u_2 \leq k$, then $U_F(u_1, u_2; \mu_1, \mu_2) = \max(\mu_1, k) = \mu_1$;
   ii. if $k < u_2$, then $U_F(u_1, u_2; \mu_1, \mu_2) = \max(\mu_1, u_2)$.

From the subcase 1 when the degree of plausibility $\mu_1 \leq k$, it can be seen that the value of the utility function U_F is either u_2 or k. When the plausible reword is greater than k then decision maker looks forward to this reword. When the plausible reword is lower than k, the decision maker focuses on threshold k. Therefore, the same conclusion is obtained as in [5]. For subcase 2 the degree of plausibility is $k < \mu_1 \leq a$, and again the conclusion is the same as in [5] (Fig. 4).

Remark 2 From the previous analysis can be concluded that the properties of the utility function U_F, obtained from a T-uninorm F with an annihilator $k \in (0, 1)$ and a continuous t-conorm S, are the same as in [5] where utility function U_F was constructed from a continuous nullnorm F with an annihilator $k \in (0, 1)$. Therefore, the utility function U_F constructed in [5] is the most general extension of a hybrid utility function from [2].

4 Conclusion

The focus of this paper is on conditional distributivity involving aggregation operators with an annihilator, namely on conditional distributivity of a continuous nullnorm or a T-uninorm in $U_{\max}$, over a continuous t-conorm. Since both a nullnorm and a T-uninorm do have an absorbing element, their impact on hybrid utility functions is investigated. In both cases a threshold that is imposed by the absorbing element is present. This property can be very useful for modelling the behavior of some decision makers, therefore further investigations will go in this direction.

References

1. Aczel, J.: Lectures on Functional Equations and Their Applications. Academic Press (1966)
2. Dubois, D., Pap, E., Prade, H.: Hybrid probabilictic-possibilistic mixtures and utility functions. In: Preferences and Decisions Under Incomplete Knowledge. Studies in Fuzziness and Soft Computing, vol. 51, pp. 51–73. Springer-Verlag (2000)
3. Grabisch, M., Marichal, J., Mesiar, R., Pap, E.: Aggregation Functions. Cambridge University Press (2009)
4. Jočić, D., Štajner-Papuga, I.: Restricted distributivity for aggregation operators with absorbing element. Fuzzy Sets Syst. **224**, 23–35 (2013)
5. Jočić, D., Štajner-Papuga, I.: Some implications of the restricted distributivity of aggregation operators with absorbing elements for utility theory. Fuzzy Sets Syst. **291**, 54–65 (2016)
6. Jočić, D., Štajner-Papuga, I.: Distributivity and conditional distributivity for T-uninorms. Inf. Sci. **424**, 91–103 (2018)
7. Klement, E.P., Mesiar, R., Pap, E.: Triangular Norms. Kluwer Academic Publishers, Dordrecht (2000)
8. Mas, M., Mesiar, R., Monserat, M., Torrens, J.: Aggregation operations with annihilator. Int. J. Gen. Syst. **34**, 1–22 (2005)

Categories of L-Fuzzy Morphological Operators on L-Fuzzy Groups and LR-Fuzzy Homomorphisms

Ksenija Varfolomejeva and Alexander Šostak

Abstract Given a completely distributive lattice L, we introduce two categories whose objects are L-powersets L^X of additive groups $(X, +, 0)$ and the morphisms are respectively forward and backward operators $R^{\to} : L^X \to L^Y$ and $R^{\leftarrow} : L^Y \to L^X$ induced by L-relations $R : X \times Y \to L$ which preserve the algebraic structure of the underlying groups $(X, +_X, 0_X)$ and $(Y, +_Y, 0_Y)$. The objects of these categories serve as the field of action for operators of fuzzy erosion and dilation. We study some properties of these operators, in particular their behaviour under operators $R^{\to}$ and $R^{\leftarrow}$.

Keywords Completely distributive lattices · L-relations · Categories of L-fuzzy groups · Morphological operators on L-fuzzy groups

1 Introduction

Mathematical morphology has its origins in geological problems centered in the processes of erosion and dilation. The founders of mathematical morphology are engineers Matheron [13] and Serra [19]. The idea of the classical mathematical morphology can be explained as the process of modifying a subset A (image) of the n-dimensional Euclidean space $\mathbb{R}^n$ by cutting out pieces of the a priory chosen S from A (in case of erosion $\mathcal{E}$) or gluing them down to A (in case of dilation $\mathcal{D}$). The set S, having "good" shape and, intuitively, small if compared with A, is called the structuring element. The process of cutting and gluing is based on the transfer of S by means of the linear structure of the underlying space $\mathbb{R}^n$. The successive application of dilation and erosion leads to the operation $\mathcal{C}$ called closing and the successive

A. Šostak
Institute of Mathematics and CS, University of Latvia, Riga, Latvia
e-mail: ksenija.varfolomejeva@lu.lv

K. Varfolomejeva (✉) · A. Šostak
Department of Mathematics, University of Latvia, Riga, Latvia
e-mail: aleksandrs.sostaks@lumii.lv

L. T. Kóczy et al. (eds.), *Computational Intelligence and Mathematics for Tackling Complex Problems 6*, Studies in Computational Intelligence 1222,
https://doi.org/10.1007/978-3-031-97879-1_7

application of erosion and dilation leads to the operation of opening $\mathcal{O}$. Opening and closing give two alternative kinds of "smoothing" the given image A. Now operations of mathematical morphology are widely and effectively used in image processing and image analysis for improving the image or highlighting its special areas. This is very important in the research of different theoretical and applied areas of science, for example, in medical diagnostics, biology, geology etc., [16, 19, 20] et al.

The above described operators of mathematical morphology are quite satisfactory when dealing with processing of crisp binary images. However, they are not applicable in case of a greyscale or colour images. The desire to solve this problem aroused interest in creating fuzzy version of mathematical morphology. As different from the classic approach, now the image A can be represented by a fuzzy subset of the space $\mathbb{R}^n$, and the structuring element S can be fuzzy as well. The linear structure of the underlying space $\mathbb{R}^n$ still remains as the base for this approach. Among the first works on fuzzy mathematical morphology were papers by B. De Baets and co-authors [7]. Later many papers on theoretical aspects of fuzzy mathematical morphology in Euclidean spaces and its applications for different practical problems were published. The paper [15] contains a detailed survey of the work done in fuzzy mathematical morphology in $\mathbb{R}^n$ in the first period of its development.

The theoretical appeal of the mathematical morphology (both crisp and fuzzy) in $\mathbb{R}^n$ as well as various practical applications is largely due to the fact that the erosion and dilation operators are related by the isotone Galois connection or adjunction principle:

Given (fuzzy)subsets A, B in $\mathbb{R}^n$ and a structural (fuzzy) set S then

$$\mathcal{D}_S(A) \leq B \Longleftrightarrow A \leq \mathcal{E}_S(B).$$

Taking this erosion-dilation relationship as the basis some researchers have shown interest in developing a general algebraic view of fuzzy mathematical morphology. Namely, they were interested in expanding the topic of fuzzy morphology into a context that would avoid the use of the linear structure of $\mathbb{R}^n$ space, but at the same time preserve "the essence" of classical mathematical morphology, that is the adjunctness of the erosion-dilation pair $(\mathcal{E}, \mathcal{D})$. The implementation of this idea began in the work [10] and was further developed in a series of works by I. Bloch, A. Heijmans, C. Ronse and other authors [2, 3] et al. In general form, it looks like this.

Given two partially ordered sets $(\mathbb{L}_1, \leq_1)$ and $(\mathbb{L}_2, \leq_1)$, a pair of mappings $\mathcal{E} : \mathbb{L}_1 \to \mathbb{L}_2$ and $\mathcal{D} : \mathbb{L}_2 \to \mathbb{L}_1$ is said to form an adjunction $(\mathcal{E}, \mathcal{D})$ if $\beta \leq_2 \mathcal{E}(\alpha) \Longleftrightarrow \mathcal{D}(\beta) \leq_1 \alpha$ for all $\alpha \in \mathbb{L}_1$, $\beta \in \mathbb{L}_2$. In this case $\mathcal{E} : \mathbb{L}_1 \to \mathbb{L}_2$ is treated as an abstract erosion and $\mathcal{D} : \mathbb{L}_2 \to \mathbb{L}_1$ is treated as an abstract dilation.

With the aim of applying a general abstract approach to mathematical morphology to more "down-to-earth" situations, some specific models of abstract mathematical morphology have been presented. For example, in [12] the general abstract approach to fuzzy mathematical morphology was implemented in the special case when $\mathbb{L}_1 = L^X$ and $\mathbb{L}_2 = L^Y$, where L is a complete lattice, X, Y are sets and operators $\mathcal{E}$ and $\mathcal{D}$ are induced by an L-relation $R : X \times Y \to L$. This approach was further developed in [21–23] et al.

In contrast to the classical (fuzzy) mathematical morphology, the advantage of the abstract approach is that it internally includes the possibility of moving from one morphological context (e.g. a lattice or a space) to another. On the other hand, the abstract approach misses the opportunity to use the linear structure of the space as well as it misses (in spite of the attempts of some authors see, e.g. [22]), the influence of the structuring element that plays the decisive role in the "classic" mathematical morphology. The main goal of our work is to develop *a categorical approach to classical fuzzy mathematical morphology*, thereby taking a step to take some of the advantages of abstract approaches to mathematical morphology. In particular, the categorical approach will permit the transition between two separate "morphological structures" that can be of interest not only theoretically but also due to some practical problems.

The structure of the paper is as follows. In Sect. 2 we give preliminary information about the tools and terminology used in the paper. In Sect. 3 we consider the category of additive groups enriched with some homomorphism-type L-relations $R : (X, +_X, 0_X) \to (Y, +_Y, 0_Y)$. The subject of Sect. 4 can be shortly described as the powerset extension of the category introduced in Sect. 3. Here we introduce two categories L-**FGROUP**$^{\to}$ and L-**FGROUP**$^{\leftarrow}$ whose objects are L-powersets of additive groups and whose morphisms are respectively forward and backward operators $R^{\to} : L^X \to L^Y$, $R^{\leftarrow} : L^Y \to L^X$ induced by L-relations $R : (X, +_X, 0_X) \to (Y, +_Y, 0_Y)$. In Sect. 5 we begin to implement the principal goal of this paper. We introduce dilation and erosion operators in the categories L-**FGROUP**$^{\to}$ and L-**FGROUP**$^{\leftarrow}$ enriched with some structured elements and study the behaviour of these operators under morphisms in the categories. The last section, "Conclusion," provides a brief summary of the results obtained in the article and outlines the directions in which we foresee the further development of this theory.

2 Preliminaries

Lattices In our work $L = (L, \leq, \wedge, \vee)$ is a complete lattice with $\mathbf{0}$ and $\mathbf{1}$ as the bottom and the top elements, see e.g. [1].

A complete lattice is called infinitely distributive or a frame, if arbitrary joins distribute over finite meets, i.e. $\left(\bigvee_{i \in I} a_i\right) \wedge b = \bigvee_{i \in I} (a_i \wedge b)$ for any $\{\alpha_i \mid i \in I\} \subseteq L$ and any $b \in L$. A lattice is called complete distributive if $\bigwedge_{i \in I} \bigvee_{j \in J_i} a_{i,j} = \bigvee_{\varphi \in \prod_{i \in I} J_i} \bigwedge_{i \in I} \alpha_{i,\varphi(i)}$ is satisfied for any family $\{\alpha_{i,j} \mid i \in I\} \subseteq L$. Obviously, a completely distributive lattice is also infinitely distributive. An important example of completely distributive lattices is complete linearly ordered lattices.

In our study we sometimes view L as a residuated lattice [14], that is a lattice $L = (L, \leq, \wedge, \vee)$ enriched with the residuum $\mapsto_{\wedge}: L \times L \to L$. Recall that a residuum on the complete lattice $L = (L, \leq, \wedge, \vee)$ is a binary operation on L connected with operation $\wedge : L \times L \to L$ by means of the Galois connection:

$$\alpha \wedge \beta \leq \gamma \Longleftrightarrow \alpha \leq \beta \mapsto_{\wedge} \gamma.$$

In particular, if L is linearly ordered, then $\alpha \mapsto_\wedge \beta = \begin{cases} 1 & \text{if } \alpha \le \beta \\ \beta & \text{otherwise} \end{cases}$.

***L*-relations** Let X and Y be sets and L C. An L-relation between sets X and Y is an L-fuzzy subset of the product $X \times Y$ that is a mapping $R : X \times Y \to L$ [25]. Let $R_1 : X \times Y \to L$ and $R_2 : Y \times Z \to L$ be L-relations. Then their composition is the L-relation $R = R_1 \circ R_2 : X \times Z \to L$ defined by $R(x, z) = \bigvee_{y \in Y} R_1(x, y) \wedge R_2(y, z)$. In case lattice L is infinitely distributive, the composition of L-relations is associative. Further, given set X let the L-relation $E_X : X \times X \to L$ be defined as $E_X(x, x) = \mathbf{1}$ for every $x \in X$ and $E_X(x, y) = \mathbf{0}$ otherwise. Then $E_X \circ R = R$ for any L-relation $R : X \times Y \to Y$ and $R \circ E_X = R$ for any L-relation $R : Y \times X \to L$. Hence sets and L-relations between them with composition defined as above form a category $\mathbf{REL}(L)$.

An L-relation is called left connected (resp. right connected) [9] if

$$\forall y \in Y \bigvee_x R(x, y) = \mathbf{1} \left(\text{resp. } \forall x \in X \bigvee_y R(x, y) = \mathbf{1}\right).$$

An L-relation is called connected if it is both left and right connected.

Proposition 1 *If the lattice L is completely distributive, then the composition of two left connected L-relations is left connected, and the composition of two right connected L-relations is right connected.*

Remark 1 The category of sets and mappings can be interpreted as a (non-full) subcategory of the category $\mathbf{REL}(L)$ as follows. Given two sets X, Y and a function $f : X \to Y$, we interpret it as the relation $F : X \times Y \to L$ defined by $F(x, y) = \begin{cases} 1 & \text{if } y = f(x) \\ 0 & \text{otherwise} \end{cases}$. Obviously, $F : X \times Y \to L$ is always right connected. It is left-connected if and only if the function $f : X \to Y$ is surjective.

Morphological operators: Dilation and erosion For the convenience of the readers and to make the paper self-contained we recall here the classical definition of the basic operators of mathematical morphology.

Let $\mathbb{R}^n$ be an n-dimensional Euclidean space and let S be a subset of $\mathbb{R}^n$, intuitively small and called the structural element. Further for every element $x \in \mathbb{R}^X$ let $\mathcal{T} : 2^X \to 2^X$ be the translation along the vector x, in particular $\mathcal{T}_x(S) = \{x' \mid x' - x \in S\}$. Now definitions of dilation $\mathcal{D}_S(A)$ and erosion $\mathcal{E}_S(A)$ of a subset $A \subset \mathbb{R}^n$ by means of the structural set S, first introduced by Serra, see e.g. [19], can be defined as follows

$$\mathcal{D}_S(A) = \{x \in X \mid \mathcal{T}_x(S) \cap A \neq \emptyset\} \quad \text{and} \quad \mathcal{E}_S(A) = \{x \in X \mid \mathcal{T}_x \subseteq A\},$$

or, explicitly, $\mathcal{D}_S(A) = \{x \mid \exists x' \in X \text{ such that } x' \in A \text{ and } x' - x \in S\}$, $\mathcal{E}_S(A) = \{x \mid \forall x' \in X \text{ if } x' - x \in S, \text{ then } x' \in A\}$. Based on these versions of definitions of dilation and erosion we define and study morphological operators on powersets of additive groups in Sect. 4.

3 Category LR-Group of Groups and LR-Homomorphisms

As the base for the study of operators of fuzzy morphology, we will use the category of groups with certain L-relations in the role of morphisms.

Dealing with groups we use the additive version. That is, we define a group as a set X equipped with a binary associative operation $+ : X \times X \to X$, and a designated identity element 0 (i.e. $0 + x = x + 0 = x$) such that every element $x \in X$ has the inverse element $-x$ (i.e. $x + (-x) = (-x) + x = 0$). We sometimes write index X in the records $+_X$ and 0_X in order to precise the support of the group.

Let $\mathcal{X} = (X, +_X, 0_X)$ and $\mathcal{Y} = (Y, +_Y, 0_Y)$ be groups and let $R : X \times Y \to L$ be an L-relation. We call L-relation $R : X \times Y \to L$ an *LR-homomorphism* from a group $(X, +_X, 0_X)$ to a group $(Y, +_Y, 0_Y)$ (in symbols $R : (X, +_X, 0_X) \to (Y, +_Y, 0_Y)$) if it is right connected and satisfies the following conditions:

1. $R(x_1 + x_2, y_1 + y_2) \geq R(x_1, y_1) \wedge R(x_2, y_2)\ \forall x_1, x_2 \in X,\ \forall y_1, y_2 \in Y$;
2. $R(x, y) = R(-x, -y)\ \forall x \in X, y \in Y$;
3. $R(0_X, 0_Y) = \mathbf{1}$.

Theorem 1 *If $R_1 : (X, +_X, 0_X) \to (Y, +_Y, 0_Y)$ and $R_2 : (Y, +_Y, 0_Y) \to (Z, +_Z, 0_Z)$ are LR-homomorphisms, where L is a completely distributive lattice, then the composition $R = R_1 \circ R_2 : (X, +_X, 0_X) \to (Z, +_Z, 0_Z)$ is an LR-homomorphism.*

Proposition 2 *Given a group $(X, +, 0)$, we define L-relation $E_X : X \times X \to L$ by setting $E(x, x') = \begin{cases} \mathbf{1} \text{ if } x = x' \\ \mathbf{0} \text{ otherwise} \end{cases}$. Then $E_X : (X, +, 0) \to (X, +, 0)$ is an LR-homomorphism. Besides, it is the identity LR-homomorphism, that is, if $R : (X, +_X, 0_X) \to (Y, +_Y, 0_Y)$ is any LR-homomorphism, then $R \circ E_X = R : X \times Y \to L$ and $E_Y \circ R = R : X \times Y \to L$.*

From Theorem 1 and Proposition 2 we get the following.

Corollary 1 *Let L be a complete completely distributive lattice. Then additive groups as objects and LR-homomorphisms as morphisms constitute a category LR-* **group**

4 Categories of L-Fuzzy Groups and LR-Fuzzy Homomorphisms

4.1 Forward and Backward Operators Induced by L-Relations

Given a complete lattice L and an L-relation $R : X \times Y \to L$, we define operators between L^X and L^Y as follows.

Definition 1

1. The upper image of a fuzzy set $A \in L^X$ under a fuzzy relation $R : X \times Y \to L$ is the fuzzy set $R^{\to}(A) \in L^Y$ defined by $R^{\to}(A)(y) = \bigvee_{x \in X}(R(x, y) \wedge A(x))$.
2. The upper preimage of a fuzzy set $B \in L^Y$ under fuzzy relation $R : X \times Y \to L$ is the fuzzy set $R^{\leftarrow}(B) \in L^X$ defined by $R^{\leftarrow}(B)(x) = \bigvee_{y \in Y}(R(x, y) \wedge B(y))$.

Considering A as a variable over L^X, we get the upper forward operator $R^{\to} : L^X \to L^Y$ induced by relation R. In turn, considering B as a variable over L^Y, we get the upper backward operator $R^{\leftarrow} : L^Y \to L^X$ induced by relation R.

Remark 2 The operators between L^X and L^Y induced by L relations similar to ours, and also defined in more general cases, in particular, when L is an arbitrary residuated lattice $(L, \wedge, \vee, *, \mapsto_*)$, were studied and used in the papers by many authors, see e.g. [11, 17, 23].

Remark 3 [18], see also [11, 17]. Let $f : X \to Y$ be a function and $F : X \times Y \to \mathbf{2} = \{0, 1\} \subseteq L$ be the corresponding relation, that is, $F(x, y) = 1 \iff y = f(x)$, and let $A \in L^X$, $B \in L^Y$. Then $F^{\to}(A)$ coincides with the "ordinary" image $f(A)$ of the fuzzy set A under mapping f and $F^{\leftarrow}(B)$ coincides with the "ordinary" preimage $f^{-1}(B)$ of the fuzzy set B under mapping f.

In the sequel we will need the connection between upper forward and backward operators established in the following lemma:

Lemma 1 *Let X, Y be sets, L a complete lattice and $R : X \times Y \to L$ a right connected L-relation. Further, let $A \in L^X$. Then $A \leq R^{\leftarrow}(R^{\to}(A))$.*

Corollary 2 *If $A \in L^X$, $B \in L^Y$ and $R : X \times Y \to L$ is a right connected L-relation such that $R^{\to}(A) \leq B$, then $A \leq R^{\leftarrow}(B)$.*

Indeed, if $R^{\to}(A) \leq B$, then $A \leq R^{\leftarrow}(R^{\to}(A)) \leq R^{\leftarrow}(B)$.

4.2 *Category L-FGROUP$^{\to}$ of L-Fuzzy Groups and $LR^{\to}$-Fuzzy Homomorphisms*

Let L be a fixed completely distributive complete lattice and let $\mathcal{X} = (X, +, 0)$ be a group. Then by an L-fuzzy group over the group $\mathcal{X}$, we call the powerset L^X. Further, let $\mathcal{Y} = (Y, +, 0)$ be another group and let $R : X \times Y \to L$ be an LR-homomorphism. Then by an $LR^{\to}$-fuzzy homomorphism induced by fuzzy relation R from the fuzzy group L^X to a fuzzy group L^Y, we call the forward operator $R^{\to} : L^X \to L^Y$.

Definition 2 L-**FGROUP**$^{\to}$ is the category whose objects are L-fuzzy groups over additive groups and whose morphisms are forward operators induced by LR-homomorphisms of the underlying groups.

We justify the correctness of this definition in the next theorem.

Theorem 2 *Let $(X, +_X, 0_X)$, $(Y, +_Y, 0_Y)$ and $(Z, +_Z, 0_Z)$ be groups and let $R_1 : X \times Y \to L$ and $R_2 : Y \times Z \to L$ be LR-homomorphisms. Then $R_1^{\to} \circ R_2^{\to} = (R_1 \circ R_2)^{\to} : L^X \to L^Z$ and $E_X^{\to} \circ R_1^{\to} = R_1^{\to}$ and $R_1^{\to} \circ E_Y^{\to} = R_1^{\to}$.*

Recalling that the composition of LR-homomorphisms is an LR-homomorphism of groups, the composition of LR-homomorphisms is associative and E_X is the identity LR-homomorphism, from Corollary 1 and from Theorem 2 we justify Definition 2. Namely, we have the following important result.

Theorem 3 *L-fuzzy groups over additive groups and $LR^{\to}$-fuzzy homomorphisms induced by LR-homomorphisms of groups constitute the category L-**FGROUP**$^{\to}$.*

4.3 *Category L-FGROUP$^{\leftarrow}$ of L-Fuzzy Groups and $LR^{\leftarrow}$-Fuzzy Homomorphisms*

Let L be a fixed completely distributive complete lattice and let $\mathcal{X} = (X, +, 0)$, $\mathcal{Y} = (Y, +, 0)$ be groups and L^X, L^Y be L-fuzzy groups over groups $\mathcal{X}$ and $\mathcal{Y}$ respectively. Further, let $R : X \times Y \to L$ be an LR-homomorphism. Then by an $LR^{\leftarrow}$-fuzzy homomorphism induced by fuzzy relation R from the fuzzy group L^Y to the fuzzy group L^X we call the backward operator $R^{\leftarrow} : L^X \to L^Y$.

Theorem 4 *Let $(X, +_X, 0_X)$, $(Y, +_Y, 0_Y)$ and $(Z, +_Z, 0_Z)$ be L-fuzzy groups and let $R_1 : X \times Y \to L$ and $R_2 : Y \times Z \to L$, be LR-homomorphisms. Then $R_1^{\leftarrow} \circ R_2^{\leftarrow} = (R_2 \circ R_1)^{\leftarrow} : L^Z \to L^X$ and $E_X^{\leftarrow} \circ R_1^{\leftarrow} = R_1^{\leftarrow} : L^Y \to L^X$ and $R_1^{\leftarrow} \circ E_Y^{\leftarrow} = R_1^{\leftarrow} : L^Y \to L^X$.*

Definition 3 L-**FGROUP**$^{\leftarrow}$ is the category whose objects are L-fuzzy groups over additive groups and whose morphisms are backward operators induced by LR-homomorphisms of the underlying groups.

The correctness of this definition is justified by the next theorem the proof of which is based on Theorem 4.

Theorem 5 *L-fuzzy groups over additive groups and $LR^{\leftarrow}$-fuzzy homomorphisms induced by LR-homomorphisms of groups constitute the category L-**FGROUP**$^{\leftarrow}$.*

5 Categories of L-Fuzzy Morphological Groups and LR-Fuzzy Homomorphisms

Dilation in the category L-GROUP$^{\to}$. Let $(X, +, 0)$ be an L-fuzzy group, and let $S \in L^X$ be an L-fuzzy set interpreted as the structural element. Further, let $A \in L^X$.

Patterned, after the "crisp" definition of dilation, see Sect. 2, we define the dilation of an L-fuzzy set A by means of the structural element S in the framework of the category of L-fuzzy groups by setting:

$$\mathcal{D}_S(A)(x) = \bigvee_{x' \in X} (A(x') \wedge S(x' - x)).$$

The proof of the following statement is obvious:

Proposition 3 *If $S \leq S'$, then $\mathcal{D}_S(A) \leq D_{S'}(A)$ for every $A \in L^X$.*

By varying A over L^X we get an operator $\mathcal{D}_S : L^X \to L^X$ called the dilation operator on the structured group $(X, S, +, 0)$. Further, let $(X, +_X, 0_X)$ and $(Y, +_Y, 0_Y)$ be groups structured by elements $S \in L^X, T \in L^Y$ respectively and let $R : X \times Y \to L$ be an LR-homomorphism such that $R^{\to}(S) \leq T$. Further let $\mathcal{D}_S : L^X \to L^X$ and $\mathcal{D}_T : L^Y \to L^Y$ be the dilation operators.

Theorem 6 *If $R : X \times Y \to L$ is an LR-fuzzy homomorphism such that $R^{\to}(S) \leq T$, then*

$$R^{\to} \circ \mathcal{D}_S \leq \mathcal{D}_T \circ R^{\to}.$$

Dilation in the category L-GROUP$^{\leftarrow}$

Theorem 7 *Let $(X, S{+}_X, 0_X)$ and $(Y, T, +_Y, 0_Y)$ be structured groups and let $R : X \times Y \to L$ be an LR-homomorphism such that $R^{\to}(S) \leq T$. Then*

$$\mathcal{D}_S \circ R^{\leftarrow} \leq R^{\leftarrow} \circ \mathcal{D}_T.$$

Erosion in the category L-GROUP$^{\leftarrow}$. Let $(X, +, 0)$ be a group, and let $S \in L^X$ be an L-fuzzy set interpreted as the structural element. Further, let $A \in L^X$. Patterned, after "crisp" definition of erosion, see Sect. 2, we define the erosion of an L-fuzzy set A by means of the structural element S in the framework of the category of L-fuzzy groups by setting:

$$\mathcal{E}_S(A)(x) = \bigwedge_{x' \in X} (A(x') \mapsto S(x' - x)).$$

By varying A over L^X we get an operator $\mathcal{E} : L^X \to L^X$ called the erosion operator on the structured group $(X, +, 0, S)$. Further, let $(X, +_X.0_X, S)$ and $(Y, +_Y, 0_Y, T)$ be groups structured by elements $S \in L^X, T \in L^Y$ respectively and let $R : X \times Y \to L$ be an LR-homomorphism such that $R^{\to}(S) \leq T$. Further let $\mathcal{E}_S : L^X \to L^X$ and $\mathcal{E}_T : L^Y \to L^Y$ be the erosion operators.

Proposition 4 *If $S \leq S'$ then $\mathcal{E}_{S'}(A) \leq \mathcal{E}_S(A)$ for every $A \in L^X$.*

Theorem 8 *Let L be a complete linearly ordered lattice and let $(X, +_X, 0_X, S)$ and $(Y, +_Y, 0_Y, T)$ be structured groups. Further let $R : X \times Y \to L$ be an LR-homomorphism such that $R^{\to}(S) \leq T$. Then $R^{\leftarrow} \circ \mathcal{E}_T \leq \mathcal{E}_S \circ R^{\leftarrow}$.*

6 Conclusion

To achieve the main goal of our work, namely to develop a categorical approach to classical fuzzy mathematical morphology, we introduce two categories L-**FGROUP**$^{\rightarrow}$ and L-**FGROUP**$^{\leftarrow}$ where L is a completely distributive lattice. The objects of these categories are L-powersets L^X of additive groups $(X, +_X, 0_X)$ and the morphisms are respectively forward and backward operators $R^{\rightarrow} : L^X \to L^Y$ and $R^{\leftarrow} : L^Y \to L^X$ induced by L-relations $R : X \times Y \to L$, which in a certain sense "respect" the algebraic structure of the underlying groups. The objects of these categories serve as the field of action for fuzzy erosion and dilation operators. We conclude the paper by studying the behaviour of these operators under morphisms of the corresponding category. Unfortunately, the restricted volume for the paper did not allow us to provide theorems and other statements with proofs as well as to present illustrative examples. We plan to do this in the extended version of this paper. This extension is currently a work in progress.

We see several current directions where the theory begun here should be continued. As one of the first problems, we see the study of properties of the categories considered here. In particular, the study of products and coproducts in these categories seems to be important, and not only theoretically, but also from the point of view of some practical applications.

Erosion and dilation in mathematical morphology serve as source operators. Their successive application leads to derived operators of opening and closing. In order to study these operators in the context of our work it seems necessary to unite the categories L-**FGROUP**$^{\rightarrow}$ and L-**FGROUP**$^{\leftarrow}$ into one category. This is a problem we are currently working on.

And last but not least. Of course, we are interested in finding some practical applications of our theory. More likely, we are expecting such applications in medical and biological problems. At least "classic" fuzzy mathematical morphology has numerous examples of application in these fields, see e.g. [4, 5].

Acknowledgements This research is part of project PID2022-139248NB-I00 funded by MCIN/AEI/10.13039/501100011033 and "ERDF A way of making Europe". The authors would like to thank the anonymous reviewers for their careful reading of the article and helpful criticism.

References

1. Birkhoff, G.: Lattice Theory, 3rd edn. AMS Providence, RI (1995)
2. Bloch, I.: Duality vs. adjunction for fuzzy mathematical morphology. Fuzzy Sets Syst. **160**, 1858–1867 (2009)
3. Bloch, I., Heijmans, H., Ronse, C.: Mathematical morphology. In: Aiello, M., Pratt-Hartmann, I., Van Benthem, J. (eds.) Handbook of Spatial Logics. Springer, pp. 857–944 (2007) (Chapter 14)
4. Boucheli, A., Pastore, J.L., Billarin, V.L.: Segmentation of medical images using fuzzy mathematical morphology. J. Comput. Sci. Technol. **7**, 256–262 (2007)

5. Caponetti, L., Castelano, G., Bastillo, M.T., Corsini, V.: Fuzzy mathematical morphology for biological images segmentation. Appl. Intell. **41**, 117–127 (2014)
6. Davey, B.A., Priestley, H.A.: Introduction to Lattices and Order, p. 298. Cambridge University Press (2002)
7. De Baets, B., Kerre, E.E., Gupta, M.: The fundamentals of fuzzy mathematical morphology part I: basic concepts. Int. J. Gen. Syst. **23**, 155–171 (1995). Part II: idempotence, convexity and cecomposition. Int. J. Gen. Syst. **23**, 307–322 (1995)
8. Gierz, G., Hofmann, K.H., Keimel, K., Lawson, J.D., Mislove, M.W., Scott, D.S.: Continuous Lattices and Domains. Cambridge University Press, Cambridge (2003)
9. Han, S.-E., Šostak, A.: On the measure of M-rough approximation of L-fuzzy sets. Soft Comput. **22**, 3843–3853 (2018)
10. Heijmans, H.J.A.M., Ronse, C.: The algebraic basis of mathematical morphology I. Dilat. Eros. Comput. Vis. Graph. Image Process. **50**(3), 245–295 (1990)
11. Klawon, F.: Fuzzy points, fuzzy relations and fuzzy functions. In: Novak, V., Perfilieva, I. (eds.) Discovering the World with Fuzzy Logic. Studies in Mathematics and Soft Computing. Cambridge University Press (2000)
12. Madrid, N., Ojeda-Aciego, M., Medina, J., Perfilieva, I.: L-fuzzy relational mathematical morphology based on adjoint triples. Inf. Sci. **474**, 75–89 (2019)
13. Matheron, G.: Random Sets and Integral Geometry. Willy (1975)
14. Morgan, W., Dilworth, R.P.: Residuated lattices. Trans. Amer. Math. Soc. **45**, 335–354 (1939)
15. Nachtegael, M., Kerre, E.E.: Classic and fuzzy approaches to mathematical morphology, Chapter 1. In: Kerre, E.E., et al. (eds.) Fuzzy Technique in Image Processing. Springer Verlag, Berlin, Heilderberg (2000)
16. Nakatsuyama, M.: Fuzzy mathematical morphology for image processing. In: ANZIS-93, pp. 75–79. Perth, Western Australia (1993)
17. Perfilieva, I., Šostak, A.: From lattice valued theories to lattice-valued analysis. In: Seising, R., et al. (eds.) Towards the Future of Fuzzy Logic, Studies in Fuzziness and Soft Computing, vol. 325. Springer, pp. 167–182 (2015) (Chapter 9)
18. Rodabaugh, S.E.: Powerset operator based foundations for poslat fuzzy set theories and topologies. Quaest. Math. **20**, 463–530 (1997)
19. Serra, J.: Image Analysis and Mathematical Morphology. Academic Press, London, New York, Paris, San Diego, San Paolo, Sydney, Tokyo, Toronto (1982)
20. Soille, P.: Morphological Image Analysis: Principles and Applications, 2nd edn. Springer Verlag (2003)
21. Šostak, A., Uļjane, I.: Manithestation of fuzzy topology in other fuzzy mathematical structures. TWMS J. Pure Appl. Math. Special Issue on Fuzzy Sets in Dealing with Imprecision and Uncertainty: Past and Future. Dedicated to the Memory of Lotfi A. Zadeh **12**, 142–172 (2021)
22. Šostak, A.: Uļjane, I: On two categories of many-valued morphological spaces. Stud. Comput. Intell. **955**, 207–217 (2018)
23. Šostak, A., Uļjane, I.: Powerset operators induced by fuzzy relations as a basis for fuzzification of various mathematical structures. Int. J. Approx. Reason. **155**, 17–39 (2023)
24. Šostak, A., Uļjane, I., Eklund, P.: Fuzzy relational mathematical morphology. Commun. Comput. Inf. Sci. (CCIS) **1239**, 712–725 (2020)
25. Zadeh, L.A.: Similarity relations and fuzzy orderings. Inf. Sci. **3**, 177–200 (1971)

Message Passing Graph Neural Network for Seeds Classification

Piotr Moszkowicz, Piotr A. Kowalski, and Tomasz Bold

Abstract Within the field of High Energy Physics various machine learning techniques are being constantly researched. Due to upcoming HL-LHC (High Luminosity Large Hadron Collider) upgrade, which aims to increase probability of occurrence of rare physical phenomenas by significant increase of number of collisions occurring each second, currently used algorithms in the area of charged particle reconstructions are not going to scale well enough. The data from tracking detectors if the LHC experiments can be represented in the form of a graph, where detected signals are represented as nodes and hypothetical connections between them are represented by edges. In particular the initial stage of track finding, the formation of seeds can be represented in that way. Within this paper we propose solution based on Graph Neural Network used to dramatically decrease number of seeds and therefore significantly reduce amount of calculations needed by downstream algorithms in order to reconstruct tracks. The numerical experiments were carried out with use a data set obtained from Monte Carlo simulation of the Open Data Detector. The initial results showed, that proposed solution based on Graph Neural Network has application potential.

Keywords Graph neural networks · Classification task · Reconstruction · Tracking · Large hadron collider · ATLAS · High energy physics

1 Introduction

The aim of Elementary Particle Physics or otherwise known as High Energy Physics (HEP) is to study the matter at subatomic level. For that the collisions at highest possible energies are being studied. These experiments are conducted with use of

P. A. Kowalski · T. Bold
Faculty of Physics and Applied Computer Science, AGH University of Krakow, Kraków, Poland

P. Moszkowicz
AGH Doctoral School, AGH University of Krakow, Kraków, Poland

P. A. Kowalski (✉)
Systems Research Institute, Polish Academy of Sciences, Warsaw, Poland
e-mail: pkowal@agh.edu.pl

L. T. Kóczy et al. (eds.), *Computational Intelligence and Mathematics for Tackling Complex Problems 6*, Studies in Computational Intelligence 1222,
https://doi.org/10.1007/978-3-031-97879-1_8

particle accelerators such as Large Hadron Collider (LHC) [1]. Since 2010 at CERN laboratory in Geneva proton beams are accelerated to the energy of several Tera-electron-Volt and collided in four instrumented areas, where particle detectors are present. In order to obtain as much data as possible, to study extremely rare processes, the frequency of these collisions has to be maximized. Nowadays the LHC is capable to deliver proton-proton collisions at rates of up to $\sim$ 40 MHz$\times$ $\sim$ 50 per second.

One of main tasks performed during the collisions' data analysis is charged particle tracking in magnetic field [2]. It allows to accurately determine position, direction, and momentum of the particles production during collision event. The area closest to the collision is filled with inner-detector, which records position of the particles moving outwards from the production point. Each record of single particle appearance is further referred to as a hit. The result of tracking algorithm should allow to identify individual trajectories of the particles. The current state-of-art solution for this problem is Combinatorial Kalman Filter (CKF) [3]. The CKF algorithm is initiated by so called seeds formation algorithm in which triplets of hits are combined. Among those seeds most of the candidates contains hits belonging to multiple particles. These so called fake seeds are considered incorrect and result in unnecessary computation needed to be performed by CKF.

To fully utilize the features of LHC machine currently High Luminosity upgrade is being planned. The upgrade will be completed by 2028 and will result in five-fold increase of collisions-per-second delivered to main experiments ATLAS [4] and CMS [5]. Currently available solutions, scale worse than quadratically with the amount of hits, which is going to increase dramatically with the increasing number of collisions per event events. Therefore a better solutions are intensively explored. The variety of proposals include usage of Quantum Annealing Optimization [6], Graph Neural Networks [7] and heterogeneous hardware architectures [8]. Our proposed solution combines usage of currently used solution (triplet seeding + CKF) and Machine Learning techniques. In order to improve tracking performance, the aim is to reduce number of fake seeds with usage of Message Passing Graph Neural Network.

The scientific novelty of this work lies in the fusion of established particle tracking methods, specifically the Combinatorial Kalman Filter, with the emerging capabilities of Graph Neural Networks. By introducing a Message Passing GNN for seed classification in the context of particle collisions, the authors offer a solution to the scalability challenges posed by the increasing collision rates at the Large Hadron Collider. Integrating machine learning techniques, particularly GNNs, into the field of Elementary Particle Physics represents a pioneering effort to optimise data analysis processes and extract meaningful insights from the vast amount of data generated by high-energy experiments.

The article is structured into six sections to provide a thorough exploration of the proposed solution. The introduction (Sect. 1) sets the stage by outlining the objectives of Elementary Particle Physics and the challenges posed by increasing collision rates. Section 2 delves into the principles and applications of Graph Neural Networks, establishing the foundation for the proposed solution. Section 3 introduces the dataset obtained from Monte Carlo simulation, emphasizing its relevance to the study. The essential section of the article (Sect. 4) details the Message Passing Graph Neural Network for seed classification, explaining the methodology and architecture. Numerical results are presented in Sect. 5, showcasing the model's performance through a confusion matrix and efficiency metrics. Finally, Sect. 6 provides a summary, consolidating the key findings and contributions of the research. The structured layout ensures a comprehensive understanding of the proposed solution and its implications in the context of High Energy Physics experiments.

2 Graph Neural Networks

Graph Neural Networks [9] are representation of newly emerging Geometric Deep Learning (GDL) [10] field, where the data is represented in non-Euclidean domain, in this case by graphs. Graph is mathematical structure, which is represented by edges (relations) between nodes (collection of entities). Each node may have certain attributes, which are represented by the feature vector. Similarly graph may have edge attributes, which are represented by edge matrix, where each row represents feature vector of single edge. Additional global context representing whole graph can also be added. With partial input of edge and/or node embeddings GNNs can output can be used to determine various features of edges, nodes or graph itself. The most popular tasks performed by GNNs are classification on graph, edge or node level and link prediction, which allows to determine the probability of existence of new edges.

The simplest GNNs uses multilayer perceptron (MLP) on each component of the graph. The MLP is applied separately for node vector, all the edges with result of pre-edge embedding and also global context vector learning embedding for entire graph. This represents single GNN layer. For classification tasks these layers can be proceeded by classification layer. The GNN performing node classification within this explanation uses only node attributes to determine the class. However sometimes the graph may contain only edge attributes, so it is impossible to do the proper classification. With usage of pooling there is possibility to combine information stored in edges or graph context and use it for node classification. In this example data from all edges connected to the single node is being gathered and aggregated via different operation. Based on pooled data the node classification task can be performed. With similar method we can perform that task on any part of the graph. The main power of GNN underlies in the concept of Message Passing, where neighboring nodes or edges are constantly exchanging information (not only during classification as previously described). At each GNN layer information from node's neigbhours and connected edges is collected and used to update its own representation. With usage of graph's connectivity it allows to capture both local and neighbouring characteristics enlarging context and allowing for additional influence from connected entities.

GNN based solutions have achieved state-of-the-art results in various tasks [11] such as molecular property prediction, networks analysis and recommendation systems. Within the social networks, GNNs can be used to predict additional links and identify influential nodes [12]. In the area of biology GNNs are found to be useful in various tasks such as Protein Design, RNA Classification or Side Effects Prediction Between Drugs among the others [13].

3 Dataset

The dataset used for this study is obtained from Monte Carlo simulation of the Open Data Detector [14] response to collision event expected at the HL-LHC era. To mimic those, about 200 elementary *pp* collisions per event were obtained using Pythia HEP generator [15]. The detector response was simulated with approximate model FATRAS [16] using tracking ACTS toolkit [17]. The number of seeds provided to the KF algorithm is of the order of $40-60 \times 10^3$. In this study seeds are categorised as genuine, if all three hits are generated by the same particle and false if the seed is constructed from hits of multiple particles. They are visualised in 3D Fig. 1. Only 1% of seeds selected at random is visualised for the sake of visibility.

Within the dataset each hit is described by its' position along Cartesian x, y, and z axis. Each seed is transformed into two edges, each edge has single attribute named *seed_quality*, which is generated by ACTS. The seed is marked as genuine one, when all three hits are caused by same particle.

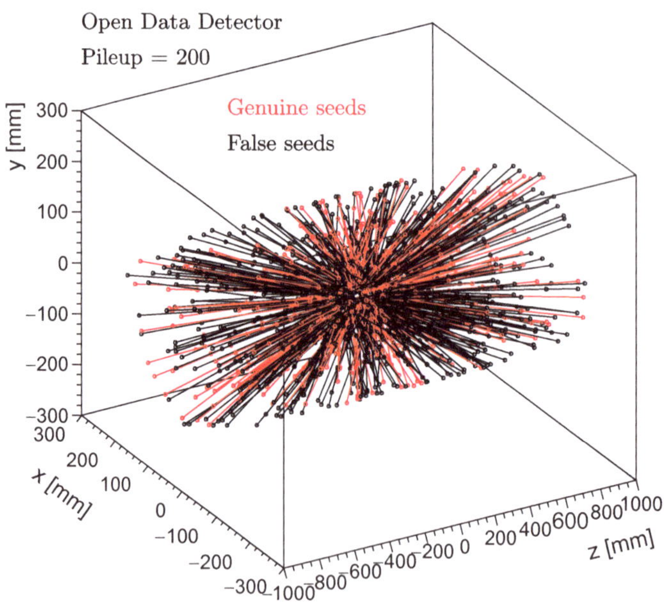

Fig. 1 Visualisation of fraction (1%) of seeds in a collision event typical for HL-LHC conditions of high pileup in open data detector. A genuine (in red) and false (black) seeds are drawn in Euclidean space

The graph is being constructed with all the edges representing seeds. Each time edge is duplicated, the one with lower $seed\_quality$ is eliminated. The resulting input graph contains hits connected by the edges, which are part of seeds. Additionally each edge has single boolean attribute describing whether it contributes to generation of genuine or false seed. The dataset contains 10,000 generated events.

4 Message Passing Graph Neural Network for Seed Classification

Proposed solution tries to determine, whether certain edge classifies as part of the genuine or false seed or not. The achieve this result input graph is being created as explained in Sect. 3.

The per-edge message function $\gamma_{E,1}$ computes four dimensional output $e'_{i,j}$ for each edge based on verticies v_i and v_j connected by the edge.

$$e'_{i,j} = \gamma_{E,1}(v_i, v_j, e_{i,j}) \tag{1}$$

where v_i and v_j are the input features of node i and j, respectively, which are the (x, y, z) coordinates of the hit. The $e_{i,j}$ is the vector representing edge embeddings containing (sq) which is the highest seed quality from the seed corresponding to the edge. These message are being aggregated by simple summation in respective neighbourhood $N(i)$ of node i.

$$m'_i = \sum_{j \in N(i)} e'_{i,j} \tag{2}$$

Steps described above constitute first cycle of message-passing. The per-edge message function ϕ generates new per-edge embeddings from its' neighbourhood. These messages are later aggregated by the sum function. There are other possible choices for the aggregation scheme—both mean and max functions are also widely used. However, the aggregations functions have to be invariant with resect to permutations inputs, and need to take any number of arguments. The sum, mean and max are example functions fulfilling this requirement.

The messages are passed to the nodes, where second update function computes new, three-dimensional embeddings for each node.

$$x'_i = \gamma_V(x_i, m'_i) \tag{3}$$

The result is the update of node features, which contains both previous attributes and also message-passed embeddings.

Finally the edge weights are computed based on new embeddings.

$$e''_{i,j} = \gamma_{E,2}(x'_i, x'_j, e'_{i,j}) \tag{4}$$

GNN's full forward pass contains update of edge embeddings, generation of the message by aggregation, computation of new verticies embeddings based on previous ones and message and finally second update of edge embeddings. The $\gamma_{E,1}$ function is MLP with ReLU activation functions on hidden layers. The $\gamma_{E,2}$ function, however contains two SAGEConv layers described as GraphSAGE by Hamilton et al. [18], each forwarded with ReLU and these forwarded with MLP. The $\gamma_{E,2}$ outputs have sigmoid activations (Fig. 2).

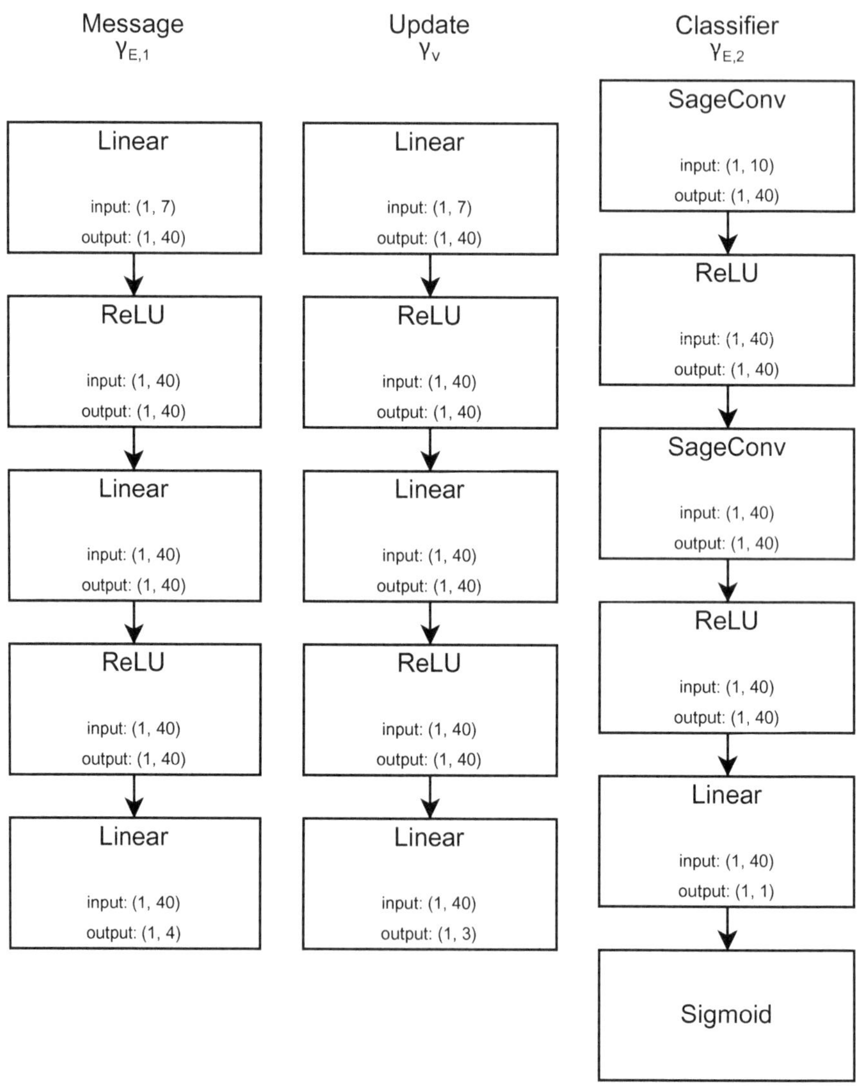

Fig. 2 Structure of the graph neural network and neural networks

The training process is optimised by the binary cross-entropy (BCE) loss between the truth targets and outputs according to the following formula:

$$-\frac{1}{N}\sum_{i=1}^{N} y_i \cdot \log(p(y_i)) + (1 - y_i) \cdot \log(1 - p(y_i)) \tag{5}$$

where actual class within the formula is represented by y_i and $\log(p(y_i))$ is the probability of that class. The result weights $e''_{i,j} \in [0, 1]$ allow to determine whether certain edge is part of the seed or not. The model was trained with 50 epochs and 40 embedding per hidden layer. The model underwent training using a 10-fold cross-validation (CV) technique, where each fold comprised distinct data sets for training, validation, and testing. Each fold consisted of 1000 events, with events from other folds utilized for testing and validation purposes.

5 Numerical Results

The network has been trained with the CV-10 folds mentioned previously. For the purpose of the numerical results, the best performing fold has been saved and re-run in order to obtain final numerical results.

The results summarising the performance of GNN are shown in Table 1. The classification accuracy of 91.34% has been achieved. Moreover, the number of True seeds marked by the network as Fake ones is just 7.49%. This metric is really important to ensure that the seeds are not lost in this process as the tracks missed due to that would never be recovered. Further studies are needed in order to lower that fraction. In particular it is planned to validate different structures of the network, try adding more per-edge embeddings (such as length between two hits) and change the contents of the messages passed through the network (Fig. 2).

Commonly used metric within High Energy Physics for evaluation of track reconstruction quality is efficiency. It is defined as following:

$$eff = \frac{|TRC|}{|TD|} \tag{6}$$

where TRC is set of True reconstructed tracks and TD is set of True tracks from dataset. For our purposes, TRC is considered to be True seeds classified by GNN,

Table 1 Confusion matrix describing seeds classification results

		Predicted	
		Fake (%)	True (%)
Real	Fake	90.65	9.35
	True	7.49	92.51

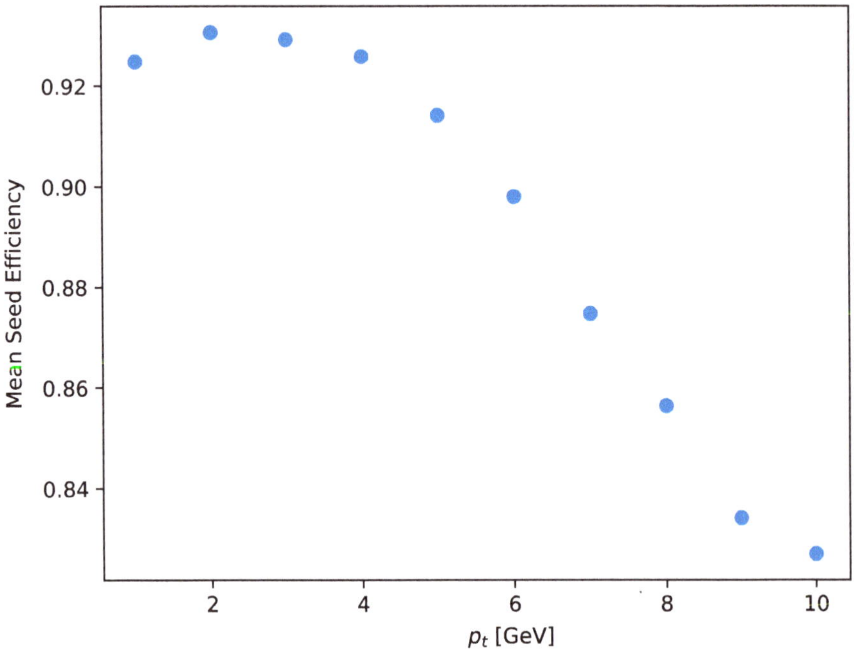

Fig. 3 Mean efficiency for various momentum values

while TD are True seeds generated from simulation. For better understanding efficiency is studied differentially as a function of various kinematic variable like particle momentum. Here each true seed has assigned transverse momentum p_T of the particle generating the track. The efficiency is studied in 9 p_T ranges of 1–10. For higher momentum the efficiency decreases, which is likely an effect caused by training the GNN on biased set of particles dominated by those of low momentum (Fig. 3).

6 Summary

This article introduces a groundbreaking approach to enhance the analysis of particle collisions at the Large Hadron Collider by integrating GNNs. Motivated by the need to improve the efficiency of particle tracking amidst the impending High Luminosity upgrade, the authors combine the traditional Combinatorial Kalman Filter with a GNN-based solution. The proposed Message Passing GNN focuses on seed classification, aiming to reduce the computational load caused by false seeds. The authors provide a comprehensive exploration of GNNs, the dataset used for training, and the numerical results demonstrating the model's efficacy. The results are promising, and the solution will be further developed. Currently planned tasks include integration of GNN usage within the ACTS toolkit to determine full tracking results. This research represents a significant step forward in addressing the evolving demands of High Energy Physics experiments.

Acknowledgements This work was partially supported by the program "Excellence initiative–âŁ"research university" for the AGH University of Krakow and by Grant for Statutory Activity from the Faculty of Physics and Applied Computer Science of the AGH University of Krakow.

References

1. Evans, L., Bryatn, P.: LHC machine. JINST **3**(S08001) (2008)
2. ATLAS Collaboration: Performance of the ATLAS track reconstruction algorithms in dense environments in LHC Run 2. JINST **77**(673) (2017)
3. Frühwirth, R.: Application of Kalman filtering to track and vertex fitting. Nucl. Instrum. Methods A **262**(444) (1987)
4. ATLAS Collaboration: The ATLAS experiment at the CERN large hadron collider. JINST **3**(S08003) (2008)
5. CMS Collaboration: The CMS experiment at the CERN LHC. EPJC **3**(S08004) (2008)
6. Zlokapa, A., Abhishek, A., Vlimant, JR., Duarte, J.M., Job J., Lidar D., Spiropulu M.: Charged particle tracking with quantum annealing optimization. Quantum Mach. Intell. **3**(27) (2021)
7. DeZoort, G., Thais, S., Duarte, J. et al.: Charged particle tracking via edge-classifying interaction networks. Comput. Softw. Big Sci. **5**(26) (2021)
8. Cerati, G., et al.: Parallelized Kalman-Filter-based reconstruction of particle tracks on many-core architectures. J. Phys. Conf. Ser. **1085**(042016) (2018)
9. Wu, Z., Pan, S., Chen, F., Long, G., Zhang, C., Yu, P.S.: A comprehensive survey on graph neural networks. IEEE Trans. Neur. Netw. Learn. Syst. **32**(1) (2021)
10. Cao, W., Yan, Z., He, Z., He, Z.: A comprehensive survey on geometric deep learning. IEEE Access **8**(35929–35949) (2020)
11. Xu, K., Hu, W., Leskovec, J., Jegelka, S.: How powerful are graph neural networks? ICLR (2019) arXiv: 1810.00826 (2019)
12. Hamilton, W., Ying, Z., Leskovec, J.: Inductive representation learning on large graphs. Adv. Neur. Inf. Process. Syst. **30** (2017)
13. Zhang, X.M., Liang, L., Liu, L., Tang, M.J.: Graph neural networks and their current applications in bioinformatics. Sec. Comput. Genom. **12** (2021)
14. Allaire, C., Gessinger, P., Hdrinka, J., Kiehn, M., Kimpel, F., Niermann, J., Salzburger, A., Sevova, S.: OpenDataDetector (2022)
15. Sjöstrand, T., Ask, S., Christiansen, J.R., Corke, R., Desai, N., Ilten, P., Mrenna, S., Prestel, S., Rasmussen, C.O., Skands, P.Z.: An introduction to PYTHIA 8.2. Comput. Phys. Commun. **191**(159–177) (2015)
16. Mechnich, J.: FATRAS—the ATLAS fast track simulation project. J. Phys. Conf. Ser. **331**(3) (2011)
17. Ai, X., Allaire, C., Calace, N., et al.: A common tracking software project. Comput. Softw. Big. Sci. **6**(8) (2022)
18. Hamilton, W.L., Ying, R., Leskovec, J.: Inductive representation learning on large graphs. arXiv: 1706.02216 (2018)

An Interval-Valued Fuzzy Soft Sets Based Decision Support Model for Route Optimization

Boldizsar Tuu-Szabo, Ruba AlMahasneh, and László T. Kóczy

Abstract Route optimization is an extensively studied NP-hard graph search problem. Many researchers applied numerous techniques to find the optimum or semi optimum solution (the one with least cost). There are many practical extensions and modifications of this problem applied using deterministic methods. However, traveling between nodes (locations) might encounter additional fuzzy cost (time) on the overall trip, whether they are traveled during the rush hour periods or if they crossed traffic regions (the city centers). Since, those factors are non-deterministic; it would be closer to reality to represent them using fuzzy numbers. In this paper, we propose a novel route optimization under road uncertainties using Interval-Valued Fuzzy Soft Sets. We use scoring technique to help determining the optimum route amongst all alternatives. Our novel approach can be looked at as a practical and closer to reality estimation for non-deterministic factors of the original abstract route optimization problem.

Keywords Routing optimization problem · Interval-valued fuzzy soft sets · Vehicle routing problem

1 Introduction

Routing optimization problem [1] has many applications in logistics and transportation. Thus, route optimization problem has always been the focus of many researchers. Choosing the best (least cost) route in transportation has become an important measure to control logistics costs and it can be applied in many areas for VRP (vehicle routing problem).

B. Tuu-Szabo (✉) · L. T. Kóczy
Department of Information Technology, Szechenyi Istvan University, Gyor, Hungary
e-mail: tuu.szabo.boldizsar@sze.hu

R. AlMahasneh · L. T. Kóczy
Telecommunications and Media Informatics, Budapest University of Technology and Economics, Budapest, Hungary

L. T. Kóczy et al. (eds.), *Computational Intelligence and Mathematics for Tackling Complex Problems 6*, Studies in Computational Intelligence 1222,
https://doi.org/10.1007/978-3-031-97879-1_9

The models which represent the connections between the nodes with constant costs have obvious limitations when simulating real world examples. It is clear that such representation is not accurate estimation for real intangible traffic circumstances. Such as traffic regions and rush hours' delays that occur during the day. Considering actual trip, clearly it is subjected not only to the topography of the given route, but also to road circumstances and conditions which affects the timing of the tour. Especially in city centers and during the rush hours, the travel times between nodes take longer, thus those two factors must be looked at as vague or uncertain (non-deterministic) values. Hence, it can be more appropriate to represent this imprecision using fuzzy numbers. This was our main motivation to introduce the novel interval-valued fuzzy soft set based model for route optimization. In this novel model, we defined two interval-valued fuzzy soft sets for representing the membership grades between the edges and jam regions, rush hour periods. This model, is one step closer towards a more efficient route optimization model in fuzzy environment.

2 Interval-Valued Fuzzy Soft Sets

The interval-valued fuzzy soft sets combine soft set theory and interval-valued fuzzy sets, which is a generalization of classical fuzzy sets initially proposed by Zadeh in 1965 [2]. Soft set theory was introduced by Molodtsov [3] in 1999, the soft set theory has been applied in several domains, such as smoothness of functions, game theory, operations research, Riemann-integration and probability. Afterwards, Maji et al. [4, 5] defined many operations for soft sets to make a detailed theoretical study on them. Also, Maji et al. successfully applied the fuzzy soft set in decision-making problems [6]. Hence, soft sets theory has been studied increasingly especially when the uncertainty is part of the problem's nature and formulation (such as the one in hand) [5, 7]. Yang et al. [8] defined the operations on fuzzy soft sets, which are based on three fuzzy logic operators: negation, triangular norm and triangular co-norm. Yang et al. [9] introduced the concept of the interval-valued fuzzy soft sets in 2009.

An interval-valued fuzzy set $\hat{X}$ on an universe U is a mapping such that $\hat{X} : U \to Int([0, 1])$, where $Int([0, 1])$ stands for the set of all closed subintervals of $[0, 1]$, the set of all interval-valued fuzzy sets on U is denoted by $\tilde{P}(U)$. Suppose that $\hat{X} \in \tilde{P}(U)$, $\forall x \in U$, $\mu_{\hat{X}}(x) = \left[\mu_{\hat{X}}^{-}(x), \mu_{\hat{X}}^{+}(x)\right]$ is called the degree of membership an element x to $\hat{X}$.

Let U be an initial universe and E be a set of parameters, a pair $\left(\tilde{F}, E\right)$ is called an interval-valued fuzzy soft set over $\tilde{P}(U)$, where $\tilde{F}$ is a mapping given by $\tilde{F} : E \to \tilde{P}(U)$.

Our model combines the interval-valued fuzzy set and fuzzy soft sets, to simulate the effect of jam regions and rush hours tangibly in any tour and eventually numerically measure their effect with respect to travelling costs, which fills a gap in the currently published researches in this domains.

3 Our Interval-Valued Fuzzy Soft Sets Based Model

We have demonstrated our model using a small size instance that consists of 8 nodes. Two congested areas (two jam regions marked with a rectangle shapes in Fig. 1) have been identified where traffic is expected to increase significantly during peak periods. In the given example, the traffic going in opposite directions was not differentiated, and only one edge was drawn between the nodes. However, if there are variations in traffic flow in both directions, a more precise model can be created by incorporating two edges.

Table 1 represents an interval-valued fuzzy soft set where U is the set of edges and A is the set of parameters, which are the two jam regions. The table shows that we do not have the precise evaluation for each object on each parameter. Instead, we have the lower and upper limits of such an evaluation. For instance, we cannot determine the exact degree of how e_{12} (between $node_1$ and $node_2$) belongs to $jam\ region_1$, but we do know that the degree is at least 0.8 and at most 1. For instance, the edge between $edge_3$ and $edge_5$ is not close to either of the jam regions. Therefore, it has a membership value of 0 for the jam regions.

Another interval-valued fuzzy soft set were defined where U is the set of edges and B is the set of parameters, which are the two, morning and afternoon peak periods ($rush\ hours_1$, $rush\ hours_2$). The interval membership grades of edges during rush hour periods vary with time. This can be explained by interval values with fuzzy membership functions, which can be determined individually for each edge based on past traffic data by experts. For instance, for some edges, the peak period is brief, while for others, an extended peak period is common. Table 2 represents the interval-valued fuzzy soft set at 8 am.

The worst traffic scenarios occur when an edge is part of a jam region during peak hours, which should be avoided if possible. These scenarios can be calculated with the intersection operation on the two interval-valued fuzzy soft sets (F, P) where U is

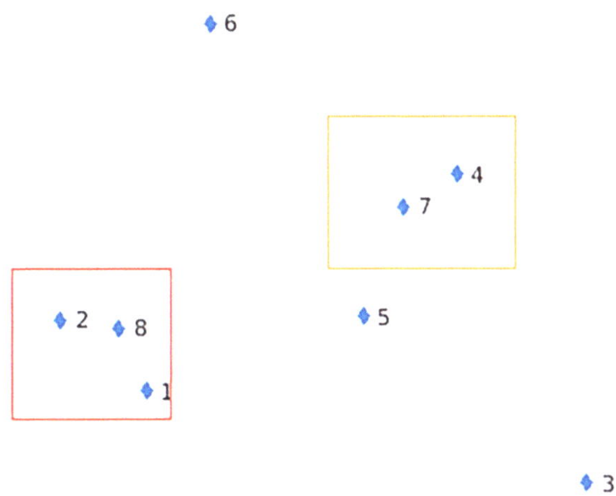

Fig. 1 A small size instance with 8 nodes and two jam regions

Table 1 The interval-values fuzzy soft set $(\tilde{F}, A)$

U	*jam region*$_1$	*jam region*$_2$
e_{12}	[0.8, 1]	[0, 0]
e_{13}	[0.1, 0.2]	[0, 0]
e_{14}	[0.1, 0.2]	[0.2, 0.3]
e_{15}	[0.1, 0.2]	[0, 0]
e_{16}	[0.25, 0.3]	[0, 0]
e_{17}	[0.1, 0.2]	[0.15, 0.25]
e_{18}	[0.8, 1]	[0, 0]
e_{23}	[0.05, 0.1]	[0, 0]
e_{24}	[0.15, 0.25]	[0.2, 0.25]
e_{25}	[0.2, 0.25]	[0, 0]
e_{26}	[0.15, 0.25]	[0, 0]
e_{27}	[0.2, 0.25]	[0.2, 0.25]
e_{28}	[0.9, 1]	[0, 0]
e_{34}	[0, 0]	[0.2, 0.3]
e_{35}	[0, 0]	[0, 0]
e_{36}	[0, 0]	[0.2, 0.25]
e_{37}	[0, 0]	[0.15, 0.25]
e_{38}	[0.05, 0.1]	[0, 0]
e_{45}	[0, 0]	[0.75, 0.85]
e_{46}	[0, 0]	[0.3, 0.4]
e_{47}	[0, 0]	[0.9, 1]
e_{48}	[0.1, 0.15]	[0.35, 0.45]
e_{56}	[0, 0]	[0.1, 0.15]
e_{57}	[0, 0]	[0.6, 0.7]
e_{58}	[0.15, 0.25]	[0, 0]
e_{67}	[0, 0]	[0.35, 0.45]
e_{68}	[0.1, 0.2]	[0, 0]
e_{78}	[0.1, 0.2]	[0.2, 0.3]

the set of the edges, P the set of the parameters P = {{*jam region*$_1$, *rush hours*$_1$}, {*jam region*$_1$, *rush hours*$_1$}, {*jam region*$_2$, *rush hours*$_1$}, {*jam region*$_2$, *rush hours*$_2$}}. The result of the intersection operation showed in Table 3.

$$\mu_{\hat{X}\cap\hat{Y}}(x) = [\inf(\mu^-_{\hat{X}}(x), \mu^-_{\hat{Y}}(x)),\ \inf(\mu^+_{\hat{X}}(x), \mu^+_{\hat{Y}}(x))] \quad (1)$$

The choice values are calculated for each edge with the following equation:

Table 2 The interval-values fuzzy soft set $(\tilde{F}, B)$

U	*rush hours*$_1$	*rush hours*$_2$
e_{12}	[0.9, 1]	[0, 0]
e_{13}	[0.5, 0.6]	[0, 0]
e_{14}	[0.7, 0.8]	[0, 0]
e_{15}	[0.6, 0.7]	[0, 0]
e_{16}	[0.6, 0.7]	[0, 0]
e_{17}	[0.7, 0.8]	[0, 0]
e_{18}	[0.9, 1]	[0, 0]
e_{23}	[0.6, 0.7]	[0, 0]
e_{24}	[0.65, 0.75]	[0, 0]
e_{25}	[0.6, 0.7]	[0, 0]
e_{26}	[0.6, 0.7]	[0, 0]
e_{27}	[0.7, 0.8]	[0, 0]
e_{28}	[0.9, 1]	[0, 0]
e_{34}	[0.65, 0.75]	[0, 0]
e_{35}	[0.4, 0.5]	[0, 0]
e_{36}	[0.6, 0.7]	[0, 0]
e_{37}	[0.6, 0.7]	[0, 0]
e_{38}	[0.55, 0.65]	[0, 0]
e_{45}	[0.7, 0.8]	[0, 0]
e_{46}	[0.6, 0.7]	[0, 0]
e_{47}	[0.9, 1]	[0, 0]
e_{48}	[0.7, 0.8]	[0, 0]
e_{56}	[0.55, 0.65]	[0, 0]
e_{57}	[0.7, 0.8]	[0, 0]
e_{58}	[0.65, 0.75]	[0, 0]
e_{67}	[0.6, 0.7]	[0, 0]
e_{68}	[0.55, 0.65]	[0, 0]
e_{78}	[0.65, 0.75]	[0, 0]

$$c_{ij} = \left[c_{ij}^-, c_{ij}^+\right] = \left[\sum_{p \in P} w_p \mu_{F(p)}^-(e_{ij}), \sum_{p \in P} w_p \mu_{F(p)}^+(e_{ij})\right] \tag{2}$$

where w_p the weights of the parameters

By utilizing weights, we can assign specific traffic scenarios with varying levels of importance. For instance, if *jam region*$_1$ is more congested than *jam region*$_2$, we can give it a higher weight. Similarly, if the morning peak hour is busier than the afternoon, we can also apply weights to enforce this priority. In the example unit weights were used. An r-score (Eq. 3) is calculated for each edge (see Table 3)

Table 3 The interval-valued fuzzy soft set calculated with intersection

U	$\{jam\ region_1, rush\ hours_1\}$	$\{jam\ region_1, rush\ hours_1\}$	$\{jam\ region_2, rush\ hours_1\}$	$\{jam\ region_2, rush\ hours_2\}$	c values	r-score
e_{12}	[0.8, 1]	[0, 0]	[0, 0]	[0, 0]	[0.8, 1]	29.09
e_{13}	[0.1, 0.2]	[0, 0]	[0, 0]	[0, 0]	[0.1, 0.2]	– 12.91
e_{14}	[0.1, 0.2]	[0, 0]	[0.2, 0.3]	[0, 0]	[0.3, 0.5]	1.09
e_{15}	[0.1, 0.2]	[0, 0]	[0, 0]	[0, 0]	[0.1, 0.2]	– 12.91
e_{16}	[0.25, 0.3]	[0, 0]	[0, 0]	[0, 0]	[0.25, 0.3]	– 5.91
e_{17}	[0.1, 0.2]	[0, 0]	[0.15, 0.25]	[0, 0]	[0.25, 0.45]	– 1.71
e_{18}	[0.8, 1]	[0, 0]	[0, 0]	[0, 0]	[0.8, 1]	29.09
e_{23}	[0.05, 0.1]	[0, 0]	[0, 0]	[0, 0]	[0.05, 0.1]	– 17.11
e_{24}	[0.15, 0.25]	[0, 0]	[0.2, 0.25]	[0, 0]	[0.35, 0.5]	2.49
e_{25}	[0.2, 0.25]	[0, 0]	[0, 0]	[0, 0]	[0.2, 0.25]	– 8.71
e_{26}	[0.15, 0.25]	[0, 0]	[0, 0]	[0, 0]	[0.15, 0.25]	– 10.11
e_{27}	[0.2, 0.25]	[0, 0]	[0.2, 0.25]	[0, 0]	[0.4, 0.5]	3.89
e_{28}	[0.9, 1]	[0, 0]	[0, 0]	[0, 0]	[0.9, 1]	31.89
e_{34}	[0, 0]	[0, 0]	[0.2, 0.3]	[0, 0]	[0.2, 0.3]	– 7.31
e_{35}	[0, 0]	[0, 0]	[0, 0]	[0, 0]	[0, 0]	– 21.31
e_{36}	[0, 0]	[0, 0]	[0.2, 0.25]	[0, 0]	[0.2, 0.25]	– 8.71
e_{37}	[0, 0]	[0, 0]	[0.15, 0.25]	[0, 0]	[0.15, 0.25]	– 10.11
e_{38}	[0.05, 0.1]	[0, 0]	[0, 0]	[0, 0]	[0.05, 0.01]	– 19.63
e_{45}	[0, 0]	[0, 0]	[0.75, 0.85]	[0, 0]	[0.7, 0.8]	20.69
e_{46}	[0, 0]	[0, 0]	[0.3, 0.4]	[0, 0]	[0.3, 0.4]	– 1.71
e_{47}	[0, 0]	[0, 0]	[0.9, 1]	[0, 0]	[0.9, 1]	31.89
e_{48}	[0.1, 0.15]	[0, 0]	[0.35, 0.45]	[0, 0]	[0.45, 0.6]	8.09
e_{56}	[0, 0]	[0, 0]	[0.1, 0.15]	[0, 0]	[0.1, 0.15]	– 14.31

(continued)

Table 3 (continued)

U	$\{jam\ region_1, rush\ hours_1\}$	$\{jam\ region_1, rush\ hours_1\}$	$\{jam\ region_2, rush\ hours_1\}$	$\{jam\ region_2, rush\ hours_2\}$	c values	r-score
e_{57}	[0, 0]	[0, 0]	[0.6, 0.7]	[0, 0]	[0.6, 0.7]	15.09
e_{58}	[0.15, 0.25]	[0, 0]	[0, 0]	[0, 0]	[0.15, 0.25]	– 10.11
e_{67}	[0, 0]	[0, 0]	[0.35, 0.45]	[0, 0]	[0.35, 0.45]	1.09
e_{68}	[0.1, 0.2]	[0, 0]	[0, 0]	[0, 0]	[0.1, 0.2]	– 12.91
e_{78}	[0.1, 0.2]	[0, 0]	[0.2, 0.3]	[0, 0]	[0.3, 0.5]	1.09

taking into account the lower and upper limits of the c-intervals, which tells us how much traffic we can expect at a particular time (in this example, at 8 am) compared to the other edges. Based on the r scores, we can determine which edges should be avoided during that time. A high r value (such as for e_{28} and e_{47}) indicates that traffic will be heavy at that time, because the edge falls into a traffic jam region with high membership grades, and one of the rush hour period membership grades is also high.

$$r_{ij} = \sum_{e_{kl} \in U} (c_{ij}^{-} - c_{kl}^{-}) + \left(c_{ij}^{+} - c_{kl}^{+}\right) \tag{3}$$

4 Conclusions

In this paper, a model based on interval-valued fuzzy soft sets was presented to describe the time and location dependence and uncertain nature of traffic. Our model can determine the road sections (edges) to avoid at a given time. In this way in the field of logistics, it can play a role in decision-making for route planning tasks. In the future, we aim to develop heuristic methods based on our decision-supporting model to effectively solve large-scale logistics route planning tasks.

References

1. Ballou, R.: Business Logistics/Supply Chain Management: Planning, Organizing and Controlling the Supply Chain. Prentice Hall Inc., New Jersey (2004)
2. Zadeh, L.A.: Fuzzy sets. Inf. Control. **8**(3), 338–353 (1965)
3. Molodtsov, D.A.: Soft set theory—first results. Comput. Math. Appl. **37**, 19–31 (1999)
4. Maji, P.K., Biswas, R., Roy, A.R.: Soft set theory. Comput. Math. Appl. **45**, 555–562 (2003)

5. Maji, P.K., Biswas, R., Roy, A.R.: Fuzzy soft sets. J. Fuzzy Math. **9**, 589–602 (2001)
6. Maji, P.K., Roy, A.R., Biswas, R.: An application of soft sets in a decision making problem. Comput. Math. Appl. **44**, 1077–1083 (2002)
7. Chen, D., Tsang, E.C.C., Yeung, D.S., Wang, X.: The parameterization reduction of soft sets and its applications. Comput. Math. Appl. **49**, 757–763 (2005)
8. Yang, X.B., Yu, D.J., Yang, J.Y., Wu, C.: Generalization of soft set theory: from crisp to fuzzy case. In: Cao, B.Y. (ed.) Proceeding of the Second International Conference on Fuzzy Information and Engineering, May, 23–16, Guangzhou, 2007. Advance on Soft Computing, vol. 40, pp. 345–354. Springer, Berlin, Heidelberg (2007)
9. Yang, X., Lin, T.Y., Yang, J., Li, Y., Yu, D.: Combination of interval-valued fuzzy set and soft set. Comput. Math. Appl. **58**(3), 521–527 (2009)

Semi-automatic Knowledge Representation and Reasoning on Vague Crime Concepts

Talissa Dreossi, Manuele Dozzi, Luca Baron, Agostino Dovier, Andrea Formisano, and Federico Costantini

Abstract In this paper we address the formal representation of vagueness in the legal domain, focusing on a "use case" in the Italian criminal law, namely the distinction of two crimes, "snatching" and "robbery". After a few epistemological clarifications on the concept of vagueness and a short premise on the legal background, we tackle the problem by adopting Answer Set Programming as modelling language. First, we encode the "static" law parts and then we enhance the encoding by learning from sentences. This is a first step for a legal reasoning system capable of evolving by doing a fully automatic learning from sentences.

Keywords Legal reasoning · Automated reasoning · ASP · Criminal law · Vagueness

1 Introduction

In legal argumentation, a specific problem arises when shaping the decision of a concrete case from a general and abstract legal provision. Namely, one has to distinguish two phenomena that are quite different: *uncertainty* and *vagueness*. Uncertainty can be appreciated from an epistemological perspective, entailing the case of a lack of information in relation to a given event or circumstance. Vagueness, instead, is defined by the fact that information is available, yet indeterminacy arises from a semantic point of view.

Consider, for example, the action performed by a thief stealing an object from a person. As we will see later, the presence of violence in the action determines the

M. Dozzi · L. Baron · F. Costantini
Department of Legal Sciences, University of Udine, Udine, Italy

T. Dreossi · A. Dovier · A. Formisano (✉)
Department of Mathematics, Computer Science and Physics, University of Udine, Udine, Italy
e-mail: andrea.formisano@uniud.it

Gruppo Nazionale per il Calcolo Scientifico, GNCS-INdAM, Rome, Italy

L. T. Kóczy et al. (eds.), *Computational Intelligence and Mathematics for Tackling Complex Problems 6*, Studies in Computational Intelligence 1222,
https://doi.org/10.1007/978-3-031-97879-1_10

type of crime committed. But establishing the presence of violence is made difficult by the vagueness of the notion of violence.

In this paper we present an introductory study that aims at addressing the problem of vagueness in the legal domain using artificial intelligence. For the sake of simplicity, we focused on a specific "use case", concerning the relationship between two offences. According to the Italian law, a street-theft—in common terms, "furto con strappo", which can be translated as "snatching"—can be included on Art. 624 bis, comma 2, of the Italian Criminal Code, and represents an aggravated form of the theft (provided by Art. 624 of the Italian Criminal Code), or it can be qualified under the Art. 628 of the Italian Criminal Code, as "rapina" (robbery, mugging) which is much more severely punished, depending on several factors that are not explicitly included in the description of the crime. The fact that these offences present a very close link, in the terms that will be described below, significantly amplifies the ambiguity that interpreters must face.

We used Answer Set Programming (ASP) as modelling language. It is a declarative language capable of addressing default negation and non-monotonic reasoning, based on the stable model semantics [14]. We encoded the relevant articles of the Italian legal code (Art. 624/"furto", Art. 624 bis/"furto con strappo", Art. 628/"rapina") as a set of ASP rules. We then used a set of decisions of the Court of Cassation to refine the model, provided the fact that this institution has the role of unifying the interpretation of law within the Italian judicial system. This second part is the encoding by hand of a "learning by examples" process. Finally, we tested the model on another set of decisions of the Court of Cassation.

The contribution is organized as follows: in Sect. 2 we briefly introduce the concept of vagueness in epistemological terms; an overview of the legal framework in which ambiguity raises and focus on the concepts that are entailed is reported in Sect. 3; in Sect. 4 we explain the methodology adopted for creating a model and testing it. The paper ends with some conclusions and possible future works.

2 About Baldness and Vagueness: An Epistemic Overview

Ordinary language often exhibits vagueness, namely a lack of clarity on whether a given object satisfies certain predicates or not. It should be noted that vagueness differs from ambiguity, which involves multiple senses of a term [19, p. 2]. Despite being distinct, vagueness and ambiguity can coexist: a given term could have multiple senses and some or all of them might be vague. Vagueness is also distinct from generality. This distinction becomes apparent if one considers a term like "prime number" which is certainly general, but not vague. That said, vagueness can sometimes constitute a means to achieve generality and adaptability, for instance, in legal contexts. A hallmark of vagueness is the presence of borderline cases [15, p. 3]. While some buildings are clearly tall (e.g., the Empire State Building) and others are not (e.g., a bungalow), many lay in between. Most analytic philosophers believe vagueness exists in language and thought, not in objects themselves; there are vague predicates but not vague objects [11]. Advocates of an ontological, i.e. not purely semantical, conception of vagueness are [8, 18, 21].

Vagueness can generate troublesome consequences such as the popular—at least among philosophers—*sorites* (Greek word for "heap") paradox. The first known version of the paradox was formulated by the Megarian philosopher Eubulides in the fourth century B.C. [23, Chap. 1]. An alternative formulation of the paradox is the following: *consider a man with no hair: undoubtedly, he is bald. However, as we incrementally add one hair at a time, based on the principle that a single hair cannot determine baldness, he remains bald: a man with 0 hair is bald, if a man with n hair is bald, so is a man with n + 1 hair, hence, a man with 1.000.000 hair is bald.* Represented using mathematical induction, if we believe as true both the base of the induction and the inductive step, we arrive at such unlikely conclusion. This presents a dilemma: either a man with no hair is not bald, which seems false, or the inductive step is inaccurate. As Boolos [3] and Sorensen [20] discussed, accepting the latter implies there is a specific hair count where one becomes bald, challenging our intuitive understanding of baldness. This means that, paradoxically, one hair could make a difference, suggesting a sharp threshold for baldness—a perspective some [23] defend despite its counter-intuitive nature. A vague predicate can be clear for some values and uncertain for others, leading to situations where it is not clear whether it holds or not, and consequently whether the statements of which it is part are true or false.

The philosophical inquiries into vagueness have implications outside the abstract domain of semantics and logic, notably within the more concrete realm of jurisprudence. Law, inherently reliant on language, is susceptible to the challenges of vagueness. Legal terms and provisions may lack clear boundaries, creating ambiguity in interpretation and application. Endicott [9] has emphasized how vagueness in legal standards, while often criticized, can be an essential feature, facilitating flexibility and adaptability to diverse situations. Waldron [22] has highlighted the potential issues of vagueness in legal contexts, especially concerning the rule of law. He underscores the need for clarity in legal norms to ensure consistent application and to uphold the ideals of fairness and justice. Both Endicott and Waldron acknowledge the presence and implications of vagueness, presenting nuanced perspectives on its role in the legal domain.

3 A Case of "Vagueness" Between "Bag-Snatching" and "Robbery" in the Italian Criminal Law

3.1 Legal Principles and Interpretation in Criminal Matters, an Overview

Current criminal law is operated by several legal principles that underpin the entire western culture, the adoption of which became accepted extensively in modern age.

Offences are encoded into crimes by legislators and modelled by tribunals whereby strict interpretation according to the principles of legality in the formulation of offences (*nullum crimen sine lege*) and in the provision of punishments (*nulla poena sine lege*), which were embodied not only into Art. 1 of the Italian Criminal Code, but also into Art. 25 of the Italian Constitution [2, 12].

From an interpretative point of view, these principles apply, in particular, when the same conduct can be referred to several crimes, resulting in what is called "concurrence of offences" (in Italian, "concorso di reati"). The concurrence of offences can be "effective" (in Italian, "reale"), when several offences are actually committed, and is distinguished in turn into "material concurrence" (in Italian, "concorso materiale di reati"), when a plurality of actions results in a plurality of offences (Articles 71–80 Italian Criminal Code) and "formal concurrence" (in Italian, "concorso formale di reati"), which occurs when one action or omission constitutes several violations of criminal law, pursuant to Art. 81 comma 1 of the Italian Criminal Code.

Different from real concurrence of offences' is "apparent concurrence", also called "fictitious concurrence", which occurs when many offences seem to be relevant, yet only one of them is applicable to the case. In this sense, the principle of substantial "*ne bis in idem*" operates, as a guarantee for convicted from being punished several times for the same material conduct. This principle is translated into Italian law by the *principle of speciality*, laid down in Art. 15 of the Italian Criminal Code, which states that "*when several criminal laws (or provisions) of the same criminal law regulate the same matter, the special law (or provision) derogates from the general one, unless otherwise stated*". In other words, the interpreter must analyse all the elements that make up the "fact", materially and historically defined (the manner in which the action or omission takes place, and their features, such as the time, or the place of the action or omission, or the quality of the victim) to select those which are relevant to the concrete case.

3.2 The Case of "Bag Snatching"

Under Italian criminal law, the basic conduct of "furto" (theft) is described in Art. 624 and can be translated as: "*Whoever takes possession of another person's movable property, capturing it from its owner, in order to gain profit for himself or others [...]*". Such crime includes several elements: (1) a movable property; (2) an action of abduction from the victim; (3) an action of taking possession by the agent; (4) the intention of profit. Theft, in this sense, does not include any kind of cooperation of the victim, who remains a mere passive subject.

Differently, "rapina" (robbery) is described in Art. 628, which can be translated as: "*Whoever, in order to procure for himself or others an unjust profit, by means of violence to the person or threat, takes possession of another person's movable property, capturing it from the person who has it [...]*". Consequently, the presence of violence—which includes any coercion exercised directly against the victim—

dictates the application of Art. 628 “rapina” instead of Art. 624 “furto” according to the principle of speciality.

To address the social alarm raised by the widespread phenomenon of “snatching”, Law No. 128 of 26 March 2001 introduced an aggravated crime of theft, provided by Art. 624 bis, comma 2, which can be translated as follows: *’Whoever takes possession of another person’s movable property, capturing it from the person holding it, in order to gain profit for himself or others, snatching it out of the person’s hand or from the person’s body [...]”*. We can observe that this offence is an intermediate figure between simple “theft”, where violence is totally absent, and “robbery”, in which violence is directed towards the victim. Despite such clear distinction, the boundaries are unclear when it comes to apply such crimes to concrete facts. Jurisprudence and scholars have deepened the distinction between “theft”, “snatching’ and “robbery” developing further criteria, which are however not definitive and inevitably leave room for vagueness. In fact, it is commonly recognised that a person’s conduct qualifies as “robbery” and not “snatching” if the item is “particularly adherent” to the victim’s body, so that the violence is no longer directed exclusively at the item, but is exercised against the individual, with the aim of overcoming its resistance [13, 17]. Even this expression, however, is vague and is variously interpreted by the courts.

4 Modelling Methodology

As mentioned, we focused on the difference between robbery and theft by snatching. To tame the vagueness of this distinction in modelling legal arguments we used Answer Set Programming (ASP) as modelling language.

ASP is a declarative rule-based language for knowledge representation and reasoning developed in the field of logic programming and non-monotonic reasoning [10]. Relevant in our context is the fact that ASP features default negation and supports non-monotonic reasoning. This makes ASP suitable for modelling problems that present different solutions (i.e., judgments) depending on how the facts/evidences regarding an event are interpreted, by resolving vagueness in different manners. Another important reason to adopt ASP is the availability of tools, such as ILASP [16], that support automated learning of ASP-rules. This will allow us to extend our ASP encoding to enable automated learning from past sentences. We will further discuss this point in Sect. 5.

Aside from the ease with which a problem can be modelled in ASP, these are the main reasons, in our opinion, for not adopting alternative logic-based formalisms, such as, for instance, many-valued or fuzzy logics. Indeed, in many-valued logics one could exploit multiple possible truth values to represent, for instance, different degrees of “closeness” to the vague notions of robbery and theft. But this would introduce different degrees of truth in the reasoning and in the solutions of the problem (the outcome of the legal argument). Instead, given the nature of the problem at hand, we are interested in solution that provide unequivocally true outcomes, leaving open the possibility of having multiple alternative solutions.

In what follows we describe the main parts of our ASP encoding.[1] Let us start by briefly introducing ASP. An ASP program is composed by:

- atoms, namely, writings of the form `p(a1, ..., an)`, where `p` is a predicate symbol and a_i are constant or variable symbols. Intuitively, such an atom states that the elements represented by `a1, ..., an` enjoy the property denoted by `p`;
- rules r of the form `H :- A1, ..., An, not B1, ..., not Bm`, where `H`, A_i and B_i are atoms, `H` is the head of r and `A1, ..., An, not B1, ..., not Bm` is its body.

A literal is an atom or the negation of an atom (e.g., `not B`). The symbol `:-` represents an implication from right to left. Rules with empty body are called facts. Rules without a head are called constraints. Intuitively, facts model properties that must be true, while constraints model conjunctions of properties that cannot be satisfied. What an ASP solver does is generating stable models, i.e. the inclusion-minimal supported sets of logical consequences of atoms and rules [14]. As there can be more than one stable model, ASP gives us more plausible, but still coherent, interpretations of the considered scenario. As an example of rule (not then used in our encoding), consider the statement "if you subtract an object owned by another person you are guilty of theft". Its encoding in ASP would be:

```
guilty_theft(R):- subtract(R, C), own(V, C), res(C)
```

where `R` is (a variable representing) the guilty of theft, `V` is the victim of theft and `C` is the subtracted object (*res*). This rule states that if the atoms `subtract(R,C)`, `own(V,C)`, and `res(C)` are true, then we can infer that `guilty_theft(R)` is true too.

As mentioned, the difference between theft by snatching and robbery lays in the exercise of violence towards someone. We describe violence as any action resulting in damage between two distinct agents (we exclude self-inflicted harm, which is not our current focus). For theft by snatching, it is essential that the object being taken has a loose physical connection to the victim. This loose adhesion is a necessary condition, because a tight connection (e.g., a watch worn on the wrist) would imply that the perpetrator would at least accept the foreseen probability of ending up in assaulting the victim. This would then qualify as a robbery.

Concerning physical adherence, the sentence n. 49832 delivered on Dec. 11, 2013 by the Section 2, Criminal Law, of the Court of Cassation (in Italian "Corte di Cassazione"), emphasized that in cases where the action is directed towards an object adhering to victim's body, the distinction between robbery and snatch theft lays in the slight or specific adhesion of the item to the person. It is from this characteristic that the presence of the intention to harm the physical integrity of the property can be inferred. This yields the following rule:

```
person_violence(R, V) :- tight_physical_adherence(V, C),
                         subtract(R, C), agent(V), R!=V.
```

[1] The complete code can be found at http://clp.dimi.uniud.it/sw/.

Within the same legal argumentation it is articulated that a conduct that from the beginning constitutes a snatch theft may, during the course of the act, turn into attempted or completed robbery. This transformation may occur when the perpetrator encounters resistance from the victim, leading to various legal classifications:

- if the perpetrator immediately ceases their conduct upon the counterparty's reaction, attempted snatch theft occurs. This is modelled as follows:

```
attempted_theft_snatch(R, V) :-
                              resistance(V, R), interruption(R).
```

- if the perpetrator attempts to overcome the resistance of another but fails and abandons the act, it constitutes an attempted robbery:

```
attempted_robbery(R, V) :- resistance(V, R), res(C),
                not interruption(R), not take_possession(R, C).
```

- if the perpetrator attempts to overcome the resistance of another and succeeds, it constitutes a completed robbery:

```
robbery(R, V) :- theft(R, V, C), person_violence(R, V).
robbery(R, V) :- theft(R, V, C), threat(R, V).
```

What has been observed so far does not seem to pose issues of vagueness, except in the determination of whether violence has actually occurred. Specifically, the definitions of tight/loose adherence appear to be highly subjective. Therefore, we have decided to further refine this concept by assigning it a level. In particular, we introduced the following rule:

```
1{adherence(S, C, L): level(L)}1 :- unknown_adherence(S, C).
```

This rule uses a syntactic extension of the base ASP language: the head of the rule contains an *aggregate literal*. Intuitively, such head is true if exactly one among the possible atoms of the form `adherence(s,c,lev)` is true (for each possible choice of values `s,c`, as determined by the predicate `unknown_adherence`). Its effect is to model non-deterministic choice among a set of alternatives. That is, for each pair `s,c`, there is exactly one value `lev` which makes the atom `adherence(s,c,lev)` true. In other words, given `s` and `c` there is only one possible level of violence `lev`. This kind of rule can be implemented via simple rewriting using simple ASP rules involving default negation [1].

A threshold can be used to propagate the non-deterministic choice:

```
tight_physical_adherence(S, C) :- adherence(S, C, L),
                                  agent(S), res(C), L>2.
loose_physical_adhesion(S, C) :-  adherence(S, C, L),
                                  agent(S), res(C), L<3.
```

This way, different assumptions are generated and propagated when the level of adherence may not be predetermined (`unknown_adherence`). Consequently, a stable model will be generated for each potential adherence level.

In the future, a similar choice could be made for the notion of violence, to better establish the severity of the penalty.

The modelling described so far does not refer to any specific law or judgment but serves as a support: a judge is still needed to determine the actual level of a particular adherence. The assistance lays in being able to have a kind of classification of adherence in past judgments, so as to place the new case in the most coherent manner possible.

5 Concluding Remarks

We focused on a pair of similar crimes (snatching vs robbery); however, our overall goal is to set the basis for repeating the process in (more) automatic way for other (groups of) crimes using a system that learns ASP rules and programs such as ILASP [16]. Since these tools do not train neural networks, but they "learn" set of rules, they need some external info (for instance, for the name of the main predicates, the number of parameters, the dependence between concepts, and so on). The idea is to encode manually the rules of the legal code, leaving the possibility of exceptions and possible future generalization or refinements. This also defines the names of interesting predicates and major dependencies. Then run the ILASP system (see e.g. [7]) on a set of benchmark judgements to produce a readable program capable either of deducing a set of possible judgements for a new case or to find formal possible explanations of judgements. Scaling issues could be handled by using parallel solvers [4–6].

In the future, we intend to proceed with the research by analyzing whether this approach is also applicable in other cases and if it can also be adopted for argumenting on a more abstract level, modelling the principle of speciality while addressing the meaning of the expression "same matter".

Furthermore, since legal interpretation is commonly an activity performed jointly by many agents (defendants, prosecutors, judges), we are aiming at including such complexity in the model, considering also the fact that judicial systems are commonly hierarchically organized, so decisions of lower courts can be overruled by higher ones. In this sense, a vague concept can be interpreted differently in the first and in the second degree of judgement, depending on the arguments that have been put forward by parties and deemed prevailing by judges.

For instance, let us imagine the scenario where, following a first degree decision—an absolution—based on absence of evidence, the victim's defendant appeals presenting new evidence (such as a video or photograph) which unequivocally demonstrates that the accused had actually committed the crime. In this example, a rule not triggered in the prior decision would be activated, resulting in a different outcome. This could be done by modelling when (whether in the first or in the second-degree) new evidences can be admitted and discussed, using constraints and default negation to address cases where a new evidence contradicts a prior one.

Acknowledgements Research partially supported by Action COST CA17124 *DigForASP*, by Inter-departmental projects on Artificial Intelligence and on Cross Border Digital Forensics (*Strategic Plan UniUD-22-25*), and by INdAM-GNCS project CUP E53C22001930001. Although the contribution is the result of a joint discussion among coauthors, Sections can be attributed as follows: 2 and 3 Manuele Dozzi, Federico Costantini and Luca Baron, 4 and 5 Talissa Dreossi, Agostino Dovier and Andrea Formisano.

References

1. Baral, C.: Knowledge Representation, Reasoning and Declarative Problem Solving. Cambridge University Press (2010)
2. Beccaria, C.: Dei delitti e delle pene. Andrea Bonducci, Firenze (1764)
3. Boolos, G.: Zooming down the slippery slope. Noûs **25**(5), 695–706 (1991)
4. Dovier, A., Formisano, A., Pontelli, E.: Parallel answer set programming. In: Hamadi, Y., Sais, L. (eds.) Handbook of Parallel Constraint Reasoning, pp. 237–282. Springer (2018). https://doi.org/10.1007/978-3-319-63516-3_7
5. Dovier, A., Formisano, A., Pontelli, E., Vella, F.: A GPU implementation of the ASP computation. In: Gavanelli, M., Reppy, J.H. (eds.) Proceedings of PADL 2016. LNCS, vol. 9585, pp. 30–47. Springer (2016), https://doi.org/10.1007/978-3-319-28228-2_3
6. Dovier, A., Formisano, A., Vella, F.: GPU-based parallelism for ASP-solving. In: Hofstedt, P., Abreu, S., John, U., Kuchen, H., Seipel, D. (eds.) Declarative Programming and Knowledge Management—DECLARE 2019. Revised Selected Papers. LNCS, vol. 12057, pp. 3–23. Springer (2019). https://doi.org/10.1007/978-3-030-46714-2_1
7. Dreossi, T.: Exploring ILASP through logic puzzles modelling. In: Dovier, A., Formisano, A. (eds.) Proceedings of CILC-23. CEUR Workshop Proceedings, vol. 3428. CEUR-WS.org (2023)
8. Dummett, M.: Is time a continuum of instants. Philosophy **75**(4), 497–515 (2000)
9. Endicott, T.A.O.: Vagueness in Law. Oxford University Press, New York (2000)
10. Erdem, E., Gelfond, M., Leone, N.: Applications of answer set programming. AI Mag. **37**(3), 53–68 (2016)
11. Evans, G.: Can there be vague objects? Analysis **38**(4), 208–208 (1978)
12. Feuerbach, P.J.A.: Lehrbuch des gemeinen in Deutschland geltenden peinlichen Rechts. Heyer, Giessen (1801)
13. Fiandaca, G., Musco, E., Cupelli, C., Masullo, M.N. (eds.): Diritto penale. Parte speciale. 2.2, I delitti contro il patrimonio. Zanichelli, Bologna, 8th edn. oCLC: 1381804647 (2023)
14. Gelfond, M., Lifschitz, V.: The stable model semantics for logic programming. In: Kowalski, R.A., Bowen, K.A. (eds.) Proceedings of ICLP/SLP, pp. 1070–1080. MIT Press (1988)
15. Hyde, D.: Vagueness. Routledge, Logic and Ontology (2008)
16. Law, M.: Inductive learning of answer set programs. Ph.D. thesis. Imperial College London (2018)
17. Ludovica, D.: Manuale di diritto penale: parte speciale II. Giuffrè Francis Lefebvre (2022)
18. Parsons, T.: Indeterminate Identity: Metaphysics and Semantics. Clarendon Press (2000)
19. Raffman, D.: Unruly Words: A Study of Vague Language. Oxford University Press, Oxford, England (2013)
20. Sorensen, R.: The sorites and the generic overgeneralization effect. Analysis **72**(3), 444–449 (2012)
21. van Inwagen, P.: How to reason about vague objects. Philos. Top. **16**(1), 255–284 (1988)
22. Waldron, J.: Vagueness in law and language: some philosophical issues. Calif. Law Rev. **82**(1), 509 (1994)
23. Williamson, T.: Vagueness. Routledge, New York (1994)

Empowering Emotional Behavior Trees with Neural Computation for Digital Forensic

Stefania Costantini, Pierangelo Dell'Acqua, Giovanni De Gasperis, and Andrea Rafanelli

Abstract Empowering interactive agents with empathetic capabilities leads, on the human side, to more trust and increased engagement. This article focuses on modeling the empathetic behavior of virtual agents interacting with humans employing behavior trees which we further enhance with neuro-symbolic capabilities. This allows the specification of various kinds of empathy for the agents to be deployed for user support in critical fields such as digital forensics. This paper is a follow-up to our work in the COST Action DigForASP.

Keywords Affective computing · Human-agent teaming · Neuro-symbolic · Behavior trees · Digital forensic

1 Introduction

Affective computing refers to the domain of computing focused on leveraging emotions, feelings, moods, and other aspects of human psychology to accomplish specific tasks [26]. Moving forward, the concept of empathy shows considerable promise

Authors Stefania Costantini, Pierangelo Dell'Acqua, Giovanni De Gasperis and Andrea Rafanelli contributed equally to this work.

S. Costantini (✉) · G. De Gasperis · A. Rafanelli
DISIM, University of L'Aquila, L'Aquila, Italy
e-mail: stefania.costantini@univaq.it

G. De Gasperis
e-mail: giovanni.degasperis@univaq.it

A. Rafanelli
e-mail: andrea.rafanelli@phd.unipi.it

P. Dell'Acqua
ITN, Linköping University, Linköping, Sweden
e-mail: pierangelo.dellacqua@liu.se

A. Rafanelli
Department Computer Science, University of Pisa, Pisa, Italy

L. T. Kóczy et al. (eds.), *Computational Intelligence and Mathematics for Tackling Complex Problems 6*, Studies in Computational Intelligence 1222,
https://doi.org/10.1007/978-3-031-97879-1_11

within affective computing research. Displaying empathy has considerable benefits in social interactions, as it fosters better relationships [2, 3, 15]. Analogously, endowing artificial agents with empathetic capabilities enhances human-agent interactions [27]. Thus, computational modeling of empathy has become an active research area with applications in education, healthcare, gaming, robotics, and more. Recent work has started investigating different aspects of empathy in human-robot interaction (c.f., [4, 20, 22]). However, achieving full empathetic competencies remains a challenge [18].

This work focuses on modeling empathetic behavior in virtual agents to be employed for user support in critical fields such as Digital Forensics as a follow-up to our work in the COST Action DigForASP [6]. To this end, in [9, 10, 12], we extended behavior trees, a modular AI formalism originally proposed for non player characters in video games. This paper proposes a further extension through a new category of nodes performing neural elaboration. Accordingly, the paper is organized as follows. Section 2 highlights some related works. Section 3 provides background on behavior trees, explaining their key capabilities and components. Section 4 introduces our proposed extension to emotionally aware behavioral trees (EABTs) for modeling neuro-symbolic computation. Section 5 demonstrates the use of our framework on a practical assistive agent scenario from the DigForASP application domain. Finally, Sect. 6 concludes the paper.

2 Related Works

Replicating empathetic behavior presents the challenge of integrating empathetic capabilities into artificial agents or computer systems, enabling them to demonstrate empathy in their actions. This challenge has been approached in various ways by researchers, who have focused on developing computational models or architectures that can simulate empathy [19]. Accordingly, many works have been developed in this area. For example, in [17] the authors propose a data-driven approach to teaching synthetic agents empathy by observing human-human social interactions. They train synthetic agents to perform tasks while manipulating their affective states to produce empathetic behaviors. In another work [24], multi-modal models for identifying opportunities for expressing empathy in human-agent interactions have been implemented. Emotional tone in language in addition to facial expressions were found to be strong indicators of dramatic sentiment in conversation that warrant an empathetic response. To automatically recognize such instances, the authors developed a multi-modal deep neural network for identifying opportunities when the agent should express positive or negative empathetic responses. The model was trained and evaluated using audio, video and language from human-agent interactions.

In [16], the authors presented a scenario where a social robot acts as a chess companion for children. They developed a framework for modeling some of the user's affective states that combines visual and task-related features. Using this model of the user, they personalized the learning environment by adapting the robot's empathetic responses to the particular preferences of the child interacting with the robot.

In [23], a method that enables a robot to augment its behavioral skills both by personally experiencing the outcomes of its own actions, as well as by interpreting the resulting rewards obtained by nearby robots and people exhibiting behaviors in its presence, has been proposed. The authors tested their method in a multi-agent setting where tasks involved actions such as pursuing a ball and striking a ball toward a target. A relevant work [1] is the Automatic Cognitive Empathy Model (ACEM). The authors want to achieve longer and more engaged human-robot interactions by considering human emotions and responding appropriately. The model continuously detected affective states based on facial expressions and generated desired empathetic behaviors adapted to the user's personality.

This work explores similar themes as the [1], since we both aim to configure empathy as a response induced by detecting human emotion. Our approach looks to external emotional stimuli to drive empathetic responses, aligning with [7] use of perceived affective states to shape suitable reactions. However, further implementation details may differ. We share the high-level goal of empathetic human-AI interaction, though build this capacity differently than existing methods. Direct comparisons could be an insightful area for future analysis.

3 Behavior Trees

Behavior trees (BTs) provide a mathematical framework for modeling an agent's execution of plans.[1] They describe the switching between a finite set of tasks in a modular way. Their strength comes from their ability to allow complex tasks to be constructed by combining simpler sub-tasks, without needing to worry about the specific implementation details of those sub-tasks [14]. Formally, a BT is a directed acyclic graph with nodes of different types. Unlike a traditional tree, nodes can have multiple parents, enabling the re-use of sub-trees. Execution proceeds from the root node downwards through the hierarchy of the tree. Each node returns one of three states during execution—success, failure, or running. A running status signals that the given node has not yet completed its designated task.

The key nodes in a BT include:

Leaf Nodes

- **Action**: An action represents a behavior that the character can perform. The action returns the state *success* or *failure* when it completes its execution, depending on the outcome. When an action needs more time to complete, it returns the state *running*. An action is depicted as a *white rounded rectangle*.
- **Condition**: A condition checks an internal or external state. An internal condition can test an external memory module and the description of empathy states. It returns either *success* or *failure*. Conditions are similar to actions except that they

[1] For details on BTs, the reader can refer to [5, 11].

execute immediately and never return *running*. A condition is represented as a *grey rounded rectangle*.

Inner Nodes

- **Sequence Selector**: A sequence selector is a node that typically has several child nodes that are executed sequentially. As long as a child node completes its execution successfully, the sequence selector continues executing the next child node in the sequence. If every child node returns success, then the sequence selector returns *success*. Should one of the child nodes return *failure*, the sequence selector immediately returns failure. If a child node returns *running*, the sequence selector also returns *running*. The sequence selector keeps track of which, if any, of its child nodes is currently running. A sequence selector is depicted as a *grey square with an arrow* across the links to its child nodes.
- **Priority Selector**: A priority selector has a list of child nodes that it tries to execute one at a time, concerning the specified order, until one of the child nodes returns *success*.
  If none of the child nodes executes successfully, the priority selector returns *failure*. If a child node returns *running*, the priority selector also returns *running*. The priority selector keeps track of which child nodes, if any, is currently running. A priority selector is represented with a *grey circle with a question mark* in it.
- **Parallel Node**: A parallel node executes all child nodes in parallel. A parallel node can have different ways of determining when to stop executing its child nodes. One may specify the number of child nodes that must execute successfully for the parallel node to return *success* and the number of child nodes that must fail for the parallel node to return *failure*. A parallel node is depicted as a *grey circle with a P* in it.
- **Decorator**: A decorator is a node that acts as a filter that places certain constraints on executing its single child node without affecting it. For instance, a decorator can prevent a child node from executing more often than, e.g., every five seconds. Decorators are represented as *diamonds with descriptive text* inside.

4 Neural Empathy-Aware Behavior Trees (NEABT)

To consider empathy and mimic human decision-making, in [9] we extended the definition of behavior trees and introduced a new type of selector called *emotional selector*, and a new type of node called *empathy node*.

Emotional Selector

The emotional selector is a node that orders its child sub-nodes based on a set of relevant decision factors and the agent's current affective state. The affective state represents the agent's emotions at a particular moment, serving as an indicator for how the *agent feels* in a human-relatable sense [10]. The underlying idea is that the agent elaborates the affective state during repeated interactions with the user and can thus tune its reaction accordingly. Once the ordering has been established by selecting the child nodes based on their probabilities, the emotional selector behaves as a priority selector. When the emotional selector has completed its execution and is executed again, the ordering of the nodes must be re-calculated. An emotional selector is represented with a *grey circle with the character E* in it.

Empathy Node
An empathy node provides an emotional evaluation of its single child node. An empathy node can only be a child of an emotional selector. This requirement is strict since empathy can only be modelled via emotional influences. Its child can be a leaf or an inner node. In the case of an inner node, its sub-tree cannot contain emotional nodes nor empathy nodes.[2] An empathy node is represented by a *dashed line circle* with the name of the empathy emotion in it.

To enable the integration of deep learning models for emotion recognition and symbolic models for planning and decision-making within emotional behavior trees, we introduced a new type of node, called *neural node*.

Neural Node
A neural node takes the current state of the environment and agent as input and makes inferences about the emotional state through a deep learning model. It contains a model such as an emotion recognition system that estimates the emotional state. These estimates are then mapped into the agent's affective state variables that parameterize the emotional selector. By continually updating the agent's internal emotional state, the neural node allows dynamic adaptation of behavior trees to the emotional context.

Neuro-Symbolic Aspects of the Framework
The integration of sub-symbolic models like neural networks with symbolic models like logic programs is referred to as *neuro-symbolic* integration. This field aims to combine the strengths of data-driven approaches that can process sensory inputs with structured knowledge representation and reasoning. There are various methods proposed for fusing these paradigms [8, 13, 25]. One technique involves encoding one module inside the structure of the other—for example, incorporating symbolic logic rules into the topology and weights of a neural network, or encoding neural network outputs as predicates within a logical reasoning framework.

In this work, we adopt the latter approach of encoding sub-symbolic processing into a symbolic structure by integrating the output of our neural model as an input node inside a behavior tree framework. The BT can then utilize those features to drive sophisticated decision making and behavior generation. Structuring the integration in this direction provides key benefits. The logical hierarchical composition of BTs can handle goals and tasks that involve planning, action sequencing, and monitoring. Meanwhile, neural networks supply the raw perceptual inputs and features.

[2] This means that emotional and empathy nodes cannot be nested among themselves.

5 Illustrative Example

In our work with the COST Action DigForASP [6], we explored methods for aiding investigators to identify suspects of crimes or misconduct by using AI and, in particular, computational logic and computational intelligence applied to digital forensics. These techniques often employ a combination of neural network-based image analysis and logical reasoning. Despite the work done, digital forensics experts still face high pressure in high-stakes cases involving serious crimes. Long hours poring through volumes of data on multiple devices, though with the help of AI tools, can take a physical and mental toll. However, the meticulous analysis of digital evidence is crucial for reconstructing events and identifying the criminals responsible.

The framework proposed in this work integrates emotional intelligence and empathy, aiming to provide effective support to investigators in high-pressure, high-stakes cases. By monitoring and analyzing the affective and physiological states of the human investigator, the intelligent personal agents can adapt their behavior to provide empathetic support, thus maintaining the productivity of the analysis without compromising the well-being of the investigators.

This section illustrates how this tool might be employed in such a scenario. The objective is to monitor human investigators' affective states and modulate the agent's behavior accordingly. The proposed solution is designed to incorporate various components, including emotional selector and empathy nodes for affective regulation. The top level of the tree is an emotional selector that chooses high-level behaviors based on the investigator's current state. Central to reading the investigator's state is an emotion detection module leveraging neural network-based techniques. This module takes in sensory data like facial expressions, vocal tones, and body language as input to the neural models. Through processing visual, auditory, and textual signals, the models provide updated estimates on relevant dimensions of the investigator's internal condition.

Specifically, we utilize neural networks trained to detect intensities of emotions based on the categorical distinctions outlined in Plutchik's model [21]. This characterizes the human affect system across six primary emotive states—*anger, fear, surprise, sadness, disgust*, and *joy*. The real-time predictions across these six emotional dimensions provide profiles of the investigator's apparent internal state. These predictions feed into emotional selectors that infer receptiveness to empathetic gestures based on the interplay between current emotions.

By formalizing the agent's behavior via an NEABT guided by affective factors, the agent can match its collaboration to the investigator's needs. This empathetic support keeps analysis productive without compromising human well-being. The approach combines emotional intelligence with forensics expertise for more effective human-machine teaming.

To illustrate our approach, below we describe the key components of our framework.[3] The model tracks both the core affective states of the human as

[3] For a discussion of the theoretical part of the framework, we point the reader to our previous work [9].

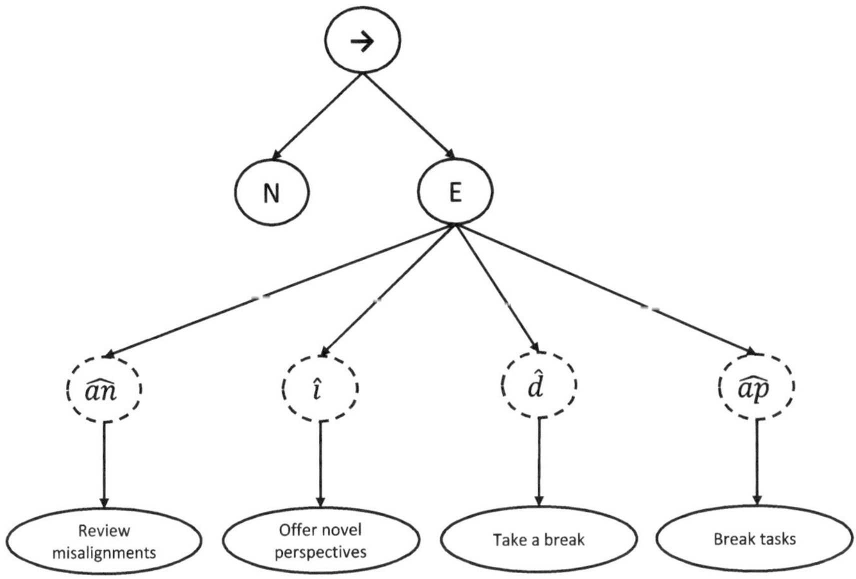

Fig. 1 NEABT of the personal assistant agent assisting digital forensics experts

well as estimated empathy states. The core affective states in our architecture are: $E = \{anger, disgust, sadness, surprise, joy, fear\}$. These are represented as continuous variables ranging from 0 to 1, with 0 indicating the absence of that emotion and 1 indicating high intensity. The empathy states are: $\widehat{E} = \{\widehat{\text{annoyance}}, \widehat{\text{interest}}, \widehat{\text{distraction}}, \widehat{\text{apprehension}}\}$. As for the affective states, the empathy states take on values between 0 and 1. For the convenience of notation, we abbreviate the empathy states as follows: $\widehat{\text{an}}$ for estimated annoyance, $\hat{\text{i}}$ for estimated interest, $\hat{\text{d}}$ for estimated distraction, and $\widehat{\text{ap}}$ for estimated apprehension.

Figure 1 shows the proposed NEABT. It is intended to be the *core* component of a personal assistant agent that incorporates empathy and emotional awareness to assist digital forensics experts in their work. At the top level is an emotional selector node that determines high-level action priorities based on the estimated affective state of the human investigator. The affective state is modeled using variables for interest, distraction, apprehension, and annoyance levels based on outputs from the integrated neural node (N). Beneath the emotional selector are branches triggering targeted responses tuned to the perceived investigator state:

- $\widehat{annoyance}$: mounting frustrations prompt a review of workflow misalignments aggravating collaboration. Validating the annoyances opens refinement dialogues to smooth interactions and avoid future impediments.
- $\widehat{interest}$: when interest levels peak, the agent offers novel visual perspectives on the case details to further stimulate investigative curiosity.

- $\widehat{distraction}$: detecting growing distraction triggers recommendations to pause the analysis and take short breaks. This allows the investigator to regain focus on the critical task elements before progress is impeded.
- $\widehat{apprehension}$: signs of uncertainty are met with suggestions to break down larger investigative tasks into smaller milestones.

6 Conclusions

The proposed NEABT architecture is intended to serve as the *core* of a personal assistant agent, incorporating empathy and emotional awareness to enhance the collaborative efforts of digital forensics experts. Its ability to adapt and respond to the investigator's emotional and physiological states offers a promising approach to human-machine teaming in digital forensics, combining emotional intelligence with forensics expertise for more effective support and collaboration.

This architecture can be used in many other application realms, for instance, to create simulations of real-world scenarios, allowing students to practice skills in a safe and controlled environment to encourage ethical, empathetic behavior.

References

1. Bagheri, E., Esteban, P.G., Cao, H., Beir, A.D., Lefeber, D., Vanderborght, B.: An autonomous cognitive empathy model responsive to users' facial emotion expressions. ACM Trans. Interact. Intell. Syst. **10**(3), 20:1–20:23 (2020). https://doi.org/10.1145/3341198
2. Barange, M., Rasendrasoa, S., Bouabdelli, M., Saunier, J., Pauchet, A.: Impact of adaptive multimodal empathic behavior on the user interaction. In: Proceedings of the 22nd ACM International Conference on Intelligent Virtual Agents, IVA '22. Association for Computing Machinery, New York, NY (2022). https://doi.org/10.1145/3514197.3549675
3. Brave, S., Nass, C., Hutchinson, K.: Computers that care: investigating the effects of orientation of emotion exhibited by an embodied computer agent. Int. J. Hum.-Comput. Stud. **62**, 161–178 (2005). https://doi.org/10.1016/j.ijhcs.2004.11.002
4. Charrier, L., Rieger, A., Galdeano, A., Cordier, A., Lefort, M., Hassas, S.: The rope scale: a measure of how empathic a robot is perceived. In: 14th ACM/IEEE International Conference on Human-Robot Interaction, HRI 2019, Daegu, South Korea, 11–14 Mar 2019, pp. 656–657. IEEE (2019). https://doi.org/10.1109/HRI.2019.8673082
5. Colledanchise, M., Ögren, P.: Behavior Trees in Robotics and AI: An Introduction. CRC Press (2018)
6. Costantini, S., De Gasperis, G., Olivieri, R.: Digital forensics and investigations meet artificial intelligence. Ann. Math. Artif. Intell. **86**(1–3), 193–229 (2019). https://doi.org/10.1007/S10472-019-09632-Y
7. Cuff, B.M., Brown, S.J., Taylor, L., Howat, D.J.: Empathy: a review of the concept. Emot. Rev. **8**(2), 144–153 (2016)
8. d'Avila Garcez, A.S., Lamb, L.C.: Neurosymbolic AI: the 3rd wave. CoRR (2020). arXiv:2012.05876

9. Dell'Acqua, P., Costantini, S., Dyoub, A., De Gasperis, G., Monaldini, A., Rafanelli, A.: Empathy-aware behavior trees for social care decision systems. In: Lessons Learned from a Multi-National Project on Developing a Platform for the Silver Economy. In: Proceeding of ProjMAN 2023—International Conference on Project MANagement, CENTERIS/ProjMAN/HCist 2023. Procedia Comput. Sci. Elsevier (2023)
10. Dell'Acqua, P., Costantini, S.: Empathetic human-agent interaction via emotional behavior trees. Intell. Artif. **17**(1), 89–100 (2023)
11. Iovino, M., Scukins, E., Styrud, J., Ögren, P., Smith, C.: A survey of behavior trees in robotics and AI. Robot. Auton. Syst. **154** (2022). https://doi.org/10.1016/j.robot.2022.104096
12. Johansson, A., Dell'Acqua, P.: Emotional behavior trees. In: 2012 IEEE Conference on Computational Intelligence and Games (CIG), pp. 355–362 (2012). https://doi.org/10.1109/CIG.2012.6374177
13. Kautz, H.A.: The third AI summer: AAAI Robert S. Engelmore memorial lecture. AI Mag. **43**(1), 93–104 (2022). https://doi.org/10.1609/AIMAG.V43I1.19122
14. Knafla, B., Champandard, A.J.: Behavior trees: introduction and memory-compact implementation. In: Algorithmic and Architectural Gaming Design: Implementation and Development, pp. 40–66. IGI Global (2012)
15. Leite, I., Castellano, G., Pereira, A., Martinho, C., Paiva, A.: Empathic robots for long-term interaction. Int. J. Soc. Robot. **6**, 329–341 (2014). https://doi.org/10.1007/s12369-014-0227-1
16. Leite, I., Pereira, A., Castellano, G., Mascarenhas, S., Martinho, C., Paiva, A.: Modelling empathy in social robotic companions. In: Ardissono, L., Kuflik, T. (eds.) Advances in User Modeling—UMAP 2011 Workshops, Girona, Spain, 11–15 July 2011, Revised Selected Papers. Lecture Notes in Computer Science, vol. 7138, pp. 135–147. Springer (2011). https://doi.org/10.1007/978-3-642-28509-7_14
17. McQuiggan, S.W., Lester, J.C.: Learning empathy: a data-driven framework for modeling empathetic companion agents. In: Nakashima, H., Wellman, M.P., Weiss, G., Stone, P. (eds.) 5th International Joint Conference on Autonomous Agents and Multiagent Systems (AAMAS 2006), Hakodate, Japan, 8–12 May 2006, pp. 961–968. ACM (2006). https://doi.org/10.1145/1160633.1160806
18. Montiel-Vázquez, E.C., Ramírez Uresti, J.A., Loyola-González, O.: An explainable artificial intelligence approach for detecting empathy in textual communication. Appl. Sci. **12**(19) (2022). https://doi.org/10.3390/app12199407. https://www.mdpi.com/2076-3417/12/19/9407
19. Paiva, A.: Empathy in social agents. Int. J. Virtual Reality **10**(11) (2015). https://doi.org/10.20870/IJVR.2011.10.1.2794
20. Park, S., Whang, M.: Empathy in human-robot interaction: designing for social robots. Int. J. Environ. Res. Public Health **19**, 1889 (2022). https://doi.org/10.3390/ijerph19031889
21. Plutchik, R.: Emotions and Life: Perspectives from Psychology, Biology, and Evolution. American Psychological Association (2003)
22. Stock-Homburg, R.: Survey of emotions in human-robot interactions: perspectives from robotic psychology on 20 years of research. Int. J. Soc. Robot. **14**(2), 389–411 (2022)
23. Takahashi, Y., Kawamata, T., Asada, M., Negrello, M.: Emulation and behavior understanding through shared values. In: 2007 IEEE/RSJ International Conference on Intelligent Robots and Systems, Sheraton Hotel and Marina, San Diego, California, USA, 29 Oct–2 Nov 2007, pp. 3950–3955. IEEE (2007). https://doi.org/10.1109/IROS.2007.4399385
24. Tavabi, L., Stefanov, K., Gilani, S.N., Traum, D.R., Soleymani, M.: Multimodal learning for identifying opportunities for empathetic responses. In: Gao, W., Meng, H.M., Turk, M.A., Fussell, S.R., Schuller, B.W., Song, Y., Yu, K. (eds.) International Conference on Multimodal Interaction, ICMI 2019, Suzhou, China, 14–18 Oct 2019, pp. 95–104. ACM (2019). https://doi.org/10.1145/3340555.3353750
25. van Bekkum, M., de Boer, M., van Harmelen, F., Meyer-Vitali, A., ten Teije, A.: Modular design patterns for hybrid learning and reasoning systems. Appl. Intell. **51**(9), 6528–6546 (2021)
26. Wang, Y., Song, W., Tao, W., Liotta, A., Yang, D., Li, X., Gao, S., Sun, Y., Ge, W., Zhang, W., Zhang, W.: A systematic review on affective computing: emotion models, databases, and

recent advances. Inf. Fusion **83–84**, 19–52 (2022). https://doi.org/10.1016/J.INFFUS.2022.03.009
27. Yalçın, Ö.N.: Evaluating empathy in artificial agents. In: 8th International Conference on Affective Computing and Intelligent Interaction (ACII), pp. 1–7. IEEE (2019)

Legal and Technical Challenges of AI in the Field of Criminal Investigations

Devika Pérez-Medina, Nicolás Madrid, and Piotr A. Kowalski

Abstract This article addresses critical challenges in the intersection of artificial intelligence, crime investigation and digital forensics, particularly in light of the proposed Regulation on Artificial Intelligence in the European Union. The focus is on mitigating biases caused by databases and algorithms, with real-world examples highlighting discriminatory biases in criminal proceedings. The article emphasises the necessity of addressing biases in both data and algorithmic decision-making to ensure fair outcomes. Another key concern explored is the lack of traceability in AI-based decisions, posing challenges to accountability and transparency, especially in the context of criminal investigations. Additionally, the article delves into the protection of private and family data in the vast datasets analysed by AI systems, referencing a legal case that underscores the potential violation of the right to a fair trial. To address this, the article proposes the development of anonymisation systems to safeguard individuals' privacy rights. The overarching theme is the need for ethical considerations and legal frameworks to guide the responsible development and deployment of AI tools in digital forensics.

D. Pérez-Medina (✉)
Department of International, Criminal and Procedure Law, University of Cádiz, Cádiz, Spain
e-mail: devika.perezmedina@gmail.com

N. Madrid
Department of Mathematics, University of Cádiz, Cádiz, Spain

P. A. Kowalski
Faculty of Physics and Applied Computer Science, AGH University of Krakow, Kraków, Poland

Systems Research Institute, Polish Academy of Sciences, Warsaw, Poland

L. T. Kóczy et al. (eds.), *Computational Intelligence and Mathematics for Tackling Complex Problems 6*, Studies in Computational Intelligence 1222,
https://doi.org/10.1007/978-3-031-97879-1_12

1 Introduction

In June 2024, Regulation (EU) 2024/1689 laying down harmonised rules on artificial intelligence (the AI Act) entered into force, establishing a comprehensive regulatory framework for the development and use of artificial intelligence within the European Union.[1] The main objective of this proposed regulation is to ensure that artificial intelligence systems that are introduced into the European market and used in the European Union are safe and, above all, respect fundamental, rights [8]. This regulation, which will be the first at a global level, comes at an extremely necessary time, considering that the use of artificial intelligence systems in areas that directly affect society is becoming more present. Criminal investigation is one of the scenarios that may be seriously affected by AI and specifically, the collection, storage and exploitation of digital evidence; the so called digital forensics analysis. This field is in continuous development with new commercial AI-based tools such as Autopsy[2] or Caine.[3] The development, distribution and use of these tools can facilitate functions, but they also run the risk of colliding with rights, including fundamental rights.

The implementation of specific legislation on AI that complements other relevant legislation in the field (such as the Data Protection Regulation [11] or the Electronic Evidence Directive [10]) aim at applying a set of rules to allow the effective development of AI tools and to ensure the respect of human rights. These regulations have different end users who will have to take into account the limits, recommendations and minimum standards for the creation, distribution and use of artificial intelligence systems, which also include the developers of these systems (mathematicians, computer scientists, data analysts, etc.). In this study, we will present three main standards that must be taken into consideration in order to make adequate legal and ethical use in the development of artificial intelligence tools in the framework of digital forensic analysis.

The consideration of these standards aim at simplifying the development of AI tools for digital forensics. Note that the development of an AI tools in Digital forensics must take into consideration all the possible inadequacies with respect to the legislation. Having standards to compare the result of an AI tool at the beginning of its development may mitigate the efforts of solving inadequacies at the last phase, where causes of inadequacy may be unsolvable. These three standards that developers should guarantee are: traceability of results, elimination of biases in the data and in the elaboration of algorithms, anonymisation of non-relevant data. The paper is structured as follows. In Sect. 2, we point out the necessity to avoid biases in the AI tools applied to criminal investigations. In this respect, we show that not only the quality of data is relevant here, but also the algorithm used for learning and the tuning done by the developer. In Sect. 3, we describe the importance of the traceability of the results obtained by an AI tool in the context of a criminal investigation. In Sect. 4, we briefly show that the development of anonymisation techniques for data used to

[1] https://eur-lex.europa.eu/legal-content/EN/TXT/?uri=CELEX%3A32024R1689.

[2] More details in https://www.autopsy.com/.

[3] More details in https://www.caine-live.net/.

infer evidences is crucial to ensure both the right of defense and the right of privacy and protection of personal data. Finally, in Sect. 5, we present some conclusions and future lines of research.

2 Elimination of Biases Caused by the Databases Used and the Ranking of Algorithms

The starting point of the problem of the use of Artificial Intelligence would be identified with the biases caused by its use. The "*European Parliament in its resolution of 6 October 2021 on artificial intelligence in criminal law and its use by law enforcement and judicial authorities in criminal matters*" pointed to the databases as the primary focus of the cause, above and beyond the algorithm or programme itself [9]. The recommendations show reasons or causes for this failure in the data used by applications based on machine learning, such as the use of historical data, outdated data, geographically defined data or biases in automation. Biases can be found not only in the database itself, but also by the algorithms used for learning and by the developer when introducing variables to be taken into account that identify ethnicities, races or specific groups, creating discrimination in the result.

This concern expressed by the parliament has also been highlighted by experts in the field. Barocas and Selbst distinguished up to five types of discriminatory biases in data mining [4]. These are: in defining variables and labeling them, collecting training data, labeling training data, feature selection and proxies. Such biases can have serious repercussions and some tools that have also been used in criminal proceedings have already been criticised because of it. One of the most known cases was the COMPAS tool as a recidivism risk assessment tool [5]. Its system determines a degree of recidivism according to proxy information; i.e., the system compares with a group of data similar to the offender the information about school grades, geography, race, age, etc. The system was used by the authorities of some US states to decide about granting parole. Studies by Angwin et al. showed how this tool is biased against people of colour, categorising them as being at higher risk than white people (up to twice as likely) whereas in practice, this percentage risk does not actually happen [2]. In the same way, Predpol, a tool for crime prediction, was widely criticised for using historical databases from the police repository to train the algorithm and generate a feedback loop [1, 7, 16]. Specifically, the use of historical data led to an increase of patrolling in social minority housing areas and, consequently, more crimes were detected; these reports were used to create a database that was used to feed the algorithm again, which created a loop of recommendation for police deployment.

Undoubtedly, the quality, choice and processing of data is fundamental to the proper use of artificial intelligence systems. These biases can come not only from using outdated historical data, but also from creating databases that lead to errors in the algorithm's decision-making process. Buolanwini and Gebru found in their research that the learning algorithms of commercial face recognition software algorithms discriminate by race and gender. The datasets used for their training contain mostly lighter-skinned subjects, which results in the fact that for black women the error rate was 34.7% while for white-skinned men the error rate was 0.8% [6].

With this exposition of challenges and problems by both, Parliament and experts in the field, the legislator incorporates this issue in Article (10) of the proposed Regulation where it refers to the obligation to subject the training, validation and test dataset, among other practices, to the examination of possible biases. This article raises a problematic issue in its Sect. 5). The legislative proposal suggests, in strictly necessary cases, the use of "special categories of data" for bias correction. In this instance, the legislator emphasizes that it will be necessary to safeguard the rights and fundamental freedoms of individuals, including technical limitations on reuse and the use of state-of-the-art security and privacy preservation measures, such as pseudonymization or encryption. These features of pseudonymization or encryption will be considered again in Sect. 4. The legislator makes a precise note of the technical limitations. Developers of these tools must therefore take this point into account in terms of data training and follow the precise indications of the Regulation. In this respect, the so-called algorithm fairness [19] is a field that explore biases appearing in algorithms and techniques to solve them. In this area there are some challenges that affect directly to criminal investigation. For example, the lack of consensus about the formal definition of algorithmic bias and fairness prevent an appropriate use by legal documents. This fact is appreciable in the AI act [8] where the use of the notion "algorithmic bias" is ambiguous. Another challenges in this area are the incorporation of causality [17] or the balance between accuracy and fairness in practical applications [14].

3 Traceability of Results

Another relevant issue is the traceability of results. As a general rule, criminal investigations are part of, or are the prelude to, criminal proceedings. Therefore, traceability is essential to guarantee fundamental rights within a prosecution (non-discrimination, fair trial, privacy, equality).

Again, the European Parliament in its resolution of 6 October 2021 [9] states that all decisions that have legal effects must be backed by a natural person who can answer for the decisions taken, so this body directly requests that the algorithms used are explainable, transparent, traceable and verifiable as a necessary part of the supervision. The literature has also pointed out that these characteristics are indispensable to guarantee the fulfilment of human rights [3, 15].

This request makes sense in view of Parliament's and experts approach that AI-based decisions should have an effective remedy, so that the subject of the judicial decision should be able to challenge the evidentiary outcome given by the AI.

This is a complicated request in some cases, given the operation of black box methods to deliver results. The main tools proposed, such as systems using biometric data—like facial recognition [12]—are almost entirely based on such algorithmic systems.

The legislator has sought to address this challenge by stating in Article 12 of the proposed AI [8] Regulation that registration capabilities shall ensure a level of traceability of the functioning of the AI system during its lifecycle.

Thus, although Article 48 of the proposal [8] states that it is the supplier that must draw up an EU declaration of conformity for each AI system in which the traceability of the system must be included, it is the technical developers who must ensure this traceability.

In the end, the legislator, an issue as transcendental as traceability, raised both by experts in the field and by the Parliament itself, does not give it sufficient development. The legislator does not define the concept of traceability and, therefore, it can be correctly interpreted how traceable the entire decision-making path from the databases to the final decision must be for the algorithm to be considered "traceable".

To the pure concept of traceability, other related issues must also be added for the purpose to be effective. It is important to ensure that artificial intelligence systems used in this field are open source. In this sense, in order to be able to trace the results, it is essential that the design of the algorithm is accessible, otherwise traceability would be impossible. Therefore, as Parliament points out [9], authorities should only acquire tools whose algorithms are open source for this type of use.

Finally, to ensure transparency, interpretability and verification, it will be essential that the algorithm design is understandable under human supervision. Therefore, the design must have a user-friendly interface and, above all, a user-friendly interface for those people who may ask for supervision or control that these tools comply with all the guarantees [20].

4 Anonymisation of Non-relevant Data

One of the main concerns in this environment is the protection of family data and respect for the fundamental right to private and family life. Machine-learning artificial intelligence systems tend to work with a huge volume of data, including personal data. If evidence extracted through digital forensic analysis techniques is analysed in a judicial process, it poses a fundamental rights challenge. On the one hand, the parties (both) must be able to access all the material used to produce the incriminating evidence in order to ensure a fair trial. This was the case Yüksel Yalçinkaya v. Türkiye in the ECtHR on 26th September 2023.[4] On this occasion, the defendant was denied access to the extraction tool and data obtained from the "Bylock" encrypted messaging software in his national trial. The ECtHR ultimately upheld the applicant's claim that Article 6 (1) of the Convention was violated.[5] The European Court

[4] More details in https://hudoc.echr.coe.int/fre?i=002-14187.

[5] Specifically, the Article 6 of European Convention on Human rights states that as "Right to a fair trial: In the determination of his civil rights and obligations or of any criminal charge against him,

of Human Rights reasoned that in order for the right to a fair trial to be guaranteed, the prosecution must disclose to the defence all material evidence in its possession, whether for or against the accused. According to the Chamber, this "material evidence" cannot be interpreted restrictively, in the sense that it cannot be limited to evidence considered relevant by the prosecution.

However, as artificial intelligence systems may work with large databases, access to them may violate the right to private and family life and the protection of personal data of those whose data is in the repository. To avoid this type of problem, developers could create anonymisation systems for data that have not been involved in the decision-making process or in the training of this tool. One technical avenue to achieve this is through advanced data anonymisation techniques. Methods such as k-anonymity, l-diversity, and t-closeness [13] provide frameworks for concealing individual identities within datasets. K-anonymity ensures that each record is indistinguishable from at least k–1 other records, thus preventing the identification of specific individuals. L-diversity expands on this concept by introducing diversity in sensitive attribute values within the anonymity sets. T-closeness, on the other hand, strives to minimise the divergence in distributions of sensitive attributes between the overall dataset and its subsets. Differential privacy, employing noise addition to data, is another robust approach to safeguard individual privacy during the analysis of large datasets [18]. The challenge with these techniques for anonymisation is that they should uphold firstly, the principles of fairness, transparency, and privacy in the development and deployment of artificial intelligence tools for digital forensics and secondly, provide a dataset capable to reproduce the same evidences gathered by an AI tool as applied to the original data.

5 Conclusions and Future Lines of Research

This article underscores the critical importance of addressing inherent challenges in the intersection of artificial intelligence and criminal investigation, particularly in the evolving regulatory landscape introduced by A Regulation of Artificial Intelligence in the European Union. The identified standards-traceability of results, elimination of biases, and anonymisation of non-relevant data-serve as foundational pillars for ethical AI development in the context of criminal investigations.

Moreover, the issue of traceability raises critical questions regarding accountability and transparency in AI-based decisions. As AI systems become integral to legal and law enforcement practices, ensuring the ability to trace and explain outcomes becomes paramount for upholding the rights of defence and effective judicial protection. The protection of private and family data presents another significant challenge, with the article highlighting the potential conflicts between the right to a fair trial and the privacy rights of individuals. The Regulation solution of anonymisation systems

everyone is entitled to a fair and public hearing within a reasonable time by an independent and impartial tribunal established by law".

seeks to strike a balance, enabling fair legal processes while safeguarding sensitive personal information.

Looking ahead, future research in this domain should explore innovative methodologies for bias detection and elimination, fostering the development of fair and accountable AI systems. Additionally, efforts should be directed towards enhancing the transparency of AI algorithms, perhaps through advancements in explainable AI, to address the complexities associated with traceability.

Furthermore, the evolving regulatory landscape necessitates ongoing interdisciplinary research involving legal scholars, ethicists, and technologists. Collaborative initiatives can contribute to the formulation of comprehensive guidelines and standards, ensuring the ethical use of AI in digital forensics. Ultimately, as the field continues to advance, it becomes imperative to prioritise ethical considerations and devise robust frameworks to navigate the intricate balance between technological innovation and fundamental rights.

Acknowledgements This article is based upon work from COST Action 17124 DigForAsp, supported by COST (European Cooperation in Science and Technology) (www.cost.eu), partially supported by the Spanish Ministry of Science and Innovation through the project PID2022-140630NB-I00, partially supported by the program "Excellence initiative—research university" for the AGH University of Krakow and by Grant for Statutory Activity from the Faculty of Physics and Applied Computer Science of the AGH University of Krakow.

References

1. Alikhademi, K., Drobina, E., Prioleau, D., Richardson, B., Purves, D., Gilbert, J.: A review of predictive policing from the perspective of fairness. Artifi. Intell. Law **30**, 1–17 (2022)
2. Angwin, J., Larson, J., Mattu, S., Kirchner, L.: Machine bias: there's software used across the country to predict future criminals. And it's biased against blacks. ProPublica (2016)
3. Bariffi, F.J.: Artificial intelligence, human rights and disability. Pensar Rev. Ciên. Jurídicas **26**(2) (2021)
4. Barocas, S., Selbst, A.D.: Big data's disparate impact. Calif. Law Rev. **104**(3), 671–732 (2016)
5. Brennan, T., Dieterich, W.: Correctional Offender Management Profiles for Alternative Sanctions (COMPAS), pp. 49–75 (2017)
6. Buolamwini, J., Gebru, T.: Gender shades: intersectional accuracy disparities in commercial gender classification. In: Friedler, S.A., Wilson, C. (eds.) Proceedings of the 1st Conference on Fairness, Accountability and Transparency. Proceedings of Machine Learning Research, vol. 81. PMLR, pp. 77–91, 23–24 Feb 2018
7. Chapman, A., Grylls, P., Ugwudike, P., Gammack, D., Ayling, J.: A data-driven analysis of the interplay between criminological theory and predictive policing algorithms. In: Proceedings of the 2022 ACM Conference on Fairness, Accountability, and Transparency, FAccT'22, pp. 36–45. Association for Computing Machinery, New York, NY, USA (2022)
8. Regulation (EU) 2024/1689 of the European Parliament and of the Council of 13 June 2024 laying down harmonised rules on artificial intelligence and amending Regulations (EC) No 300/2008, (EU) No 167/2013, (EU) No 168/2013, (EU) 2018/858, (EU) 2018/1139 and (EU) 2019/2144 and Directives 2014/90/EU, (EU) 2016/797 and (EU) 2020/1828
9. European Parliament: Resolution on artificial intelligence in criminal law and its use by the police and judicial authorities in criminal matters, 06 Oct 2021

10. European Parliament and the Council of European Union: Directive (EU) 2023/1544 of the European Parliament and of the Council of 12 July 2023 laying down harmonised rules on the designation of designated establishments and the appointment of legal representatives for the purpose of gathering electronic evidence in criminal proceedings
11. European Parliament and the Council of European Union: Regulation (EU) 2016/679 of the European Parliament and of the Council (General Protect Data Regulation)
12. Kasar, M.M., Bhattacharyya, D., Kim, T.: Face recognition using neural network: a review. Int. J. Secur. Appl. **10**(3), 81–100 (2016)
13. Li, N., Li, T., Venkatasubramanian, S.: t-closeness: privacy beyond k-anonymity and l-diversity. In: 2007 IEEE 23rd International Conference on Data Engineering, pp. 106–115 (2007)
14. Liu, S., Vicente, L.: Accuracy and fairness trade-offs in machine learning: a stochastic multi-objective approach (2020)
15. Mirzazadeh, I.: Artificial intelligence (AI) and violation of human rights. Available at SSRN 4310188 (2022)
16. Mohler, G., Short, M., Malinowski, S., Johnson, M., Tita, G., Bertozzi, A., Brantingham, P.: Randomized controlled field trials of predictive policing. J. Am. Stat. Assoc. **110** (2015)
17. Pearl, J.: Causality: Models, Reasoning and Inference, 2nd edn. Cambridge University Press, USA (2009)
18. Tu, Z., Zhao, K., Xu, F., Li, Y., Su, L., Jin, D.: Protecting trajectory from semantic attack considering k-anonymity, l-diversity, and t-closeness. IEEE Trans. Netw. Serv. Manag. **16**(1), 264–278 (2019)
19. Wang, X., Zhang, Y., Zhu, R.: A brief review on algorithmic fairness. Manag. Syst. Eng. **1**(1), 7 (2022)
20. Yeung, K., Howes, A., Pogrebna, G.: AI governance by human rights-centered design, deliberation, and oversight. In: The Oxford Handbook of Ethics of AI, pp. 77–106 (2020)

Fingerprint Revolution: Unleashing the Potential of Modified Bacterial Memetic Evolution for a Paradigm Shift in Fingerprint Recognition and Optimization

Ahmad A. Momani and László T. Kóczy

Abstract Biometrics refers to the science of measuring and analyzing biological or behavioral characteristics to identify people. It has been used in many fields including digital forensics, identification systems, and security. The most common biometric nowadays is fingerprint, which has improved significantly in the last two decades. Several issues in fingerprint identification and authentication systems have been raised due to factors such as displacement of the finger while scanning, fingerprint rotation, major cuts, distortions, and substandard quality images lowering the overall system efficiency. Evolutionary algorithms evolved in recent years to enhance the system performance over time. The proposed system uses a modified version of the bacterial memetic evolutionary algorithm to overcome the identification issues and to help and support forensic experts to make reliable decisions faster. The proposed system was evaluated on five different databases and the results demonstrated that the system succeeded in identifying the correct match from the first candidate in all cases among all examined databases.

Keywords Biometrics · Fingerprint · Evolutionary algorithms · Bacterial algorithm · Memetic algorithm

1 Introduction

Biometrics is an important field of study that is considered an interdisciplinary field combining biology and authentication schemes based on biological characteristics. The key aspect of biometrics is to reach the highest levels of Information security,

---

A. A. Momani (✉) · L. T. Kóczy
Department of Information Technology, Széchenyi István University, Győr, Hungary
e-mail: momani.ahmad@sze.hu

L. T. Kóczy
e-mail: koczy@sze.hu

L. T. Kóczy et al. (eds.), *Computational Intelligence and Mathematics for Tackling Complex Problems 6*, Studies in Computational Intelligence 1222,
https://doi.org/10.1007/978-3-031-97879-1_13

encompassing confidentiality, integrity, and information availability. It has unlimited support for security measures such as user identification systems, user authentication, and non-repudiation, especially in systems that use passwords, PINs, and tokens. Two main types of biometric traits are physiological (body shape-related) such as user fingerprints, face recognition, hand shape, and ear characteristics, while the behavioral (related to actions) includes traits like keystroke dynamics which analyze the user typing rhythm [1, 2], and voice patterns.

Biometric authentication is gaining importance in various computer-based applications due to the demand for secure data. It plays an essential role in e-commerce, online as well as offline standalone systems, and forensic science. The application of biometric authentication in forensic science is to identify individuals and verify their identity in the context of criminal investigations and legal proceedings [2, 3].

Optimization algorithms are used to refine a system, surpassing alternative methods that exist in the same field. It requires to present different sets of inputs to derive their corresponding outputs. The main goal of optimization is to identify a specific point or a set of points within the search space that represents the best or optimal solution. One example is the Genetic Algorithm (GA) which is represented by selection, crossover, and mutation processes inspired by the biological evolution (in genetics) and the natural selection theory. Genetic algorithms are utilized in search optimization problems in the field of machine learning and computational intelligence [4]. Another example is the Bacterial Evolutionary Algorithm (BEA) which is influenced by the transduction process, where the bacteria can transfer DNA during mating. This mechanism facilitates the spread of specific bacterial traits across the entire population. Hence, this mechanism of genetic recombination shapes the process of microbial evolution [5].

Despite the continuous advancements in fingerprint identification systems over the past two decades, certain challenging domains require further improvement. Traditional image processing methods encounter difficulties in handling fingerprints with significant cuts and distortions. In some cases, the fingerprint core may be unclear or nonexistent. Consequently, these challenges lead to a significant degradation in system efficiency and performance. Thus, there is a pressing need to explore alternative solutions, such as optimization algorithms, to address these issues. Using modified optimization algorithms in the fingerprint identification and recognition field will eliminate the need to extract fingerprint minutiae (Minutiae-based methods) or image details (Image-based methods), instead, a modified optimization of all impressions for each individual in the database can be used to find the best match accurately. This study introduces a novel approach to enhance fingerprint-matching outcomes through the application of modified optimization algorithms. Additionally, innovative methodologies are suggested for processing fingerprint images, aiming to eliminate undesired elements of the fingerprint image during the matching process and reduce the algorithm complexity time by dividing the fingerprint image into three segments to improve the efficiency of the algorithm. Finally, a curated ordered list of three potential matches will be presented to digital forensics experts for further analysis. The list of candidates will be presented in an intelligible format, considering the uncertainty levels determined by the proposed system.

In summary, this research paper introduces the following key contributions:

1. We introduce a novel approach to eliminate undesired elements from fingerprint images, particularly those that can significantly influence matching results by being included in the matching process (such as the white spaces surrounding the image).
2. We introduce the concept of dividing the image into three distinct segments: top, middle, and bottom for further processing, to enhance both matching accuracy and efficiency.
3. We propose an innovative matching method that incorporates the concepts of bacterial translocation mutation and optimal chromosome selection processes for fingerprint impressions belonging to the same individual in the database.

The structure of the paper is as follows: Sect. 2 offers a detailed description of the proposed system, encompassing fingerprint image key contributions through to the matching process. Section 3 presents the outcomes of tests conducted on the FVC 2000, 2002, and 2004 databases using the MATLAB software program. Lastly, Sect. 4 concludes the paper based on the results obtained from the system.

2 BioPrintGuard: Modified Bacterial Algorithm Fingerprint Identification System (MBAFIS) Overview

The proposed fingerprint identification system, MBAFIS, has five stages implemented using the MATLAB software program. In this paper, we will employ a selection of our initial methods created in [3, 6]. The proposed approach is discussed as follows:

2.1 Enhancing and Binarizing Fingerprint Images

Fingerprint enhancement strives to enhance the visibility and quality of the images by addressing noise and distortions of fingerprints. The process begins with Fourier domain analysis, as proposed by Chikkerur et al. [7], followed by the conversion of the image to binary codes. Subsequently, an additional enhancement layer proposed in our previous work [6] is applied to further reduce the noise of the enhanced images.

2.2 Rotation Solver Method

In the field of fingerprint identification systems, the conventional approaches provided solutions to the challenges posed by rotated images. In this study, we adopt the innovative rotation solver method, previously introduced in our research [3],

Table 1 Removing white spaces around fingerprint Images

FVC 2002 (102_5)	
FVC 2004 (102_1)	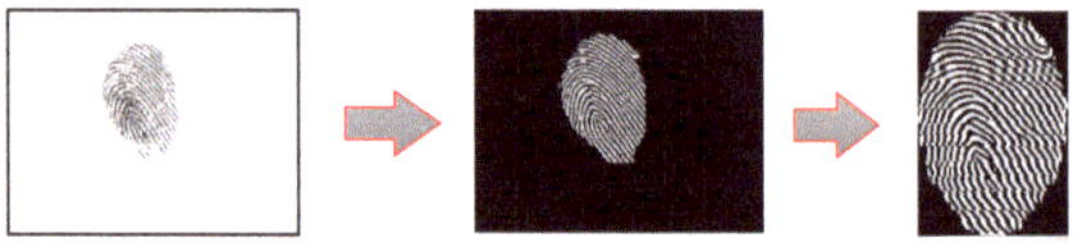

which has demonstrated efficacy in terms of time efficiency, diminished computational complexity, and consistent image resolution regardless of the rotation angle.

2.3 *Remove the White Spaces Around the Fingerprint Image*

The subsequent step in preparing the fingerprint image for matching involves the removal of white spaces surrounding it. This step is crucial as it significantly impacts the similarity results, especially in cases where fingerprints have cuts or lack a core.

To mitigate these issues, we have developed a new function designed to trim unwanted white spaces from the top, bottom, left, and right sides of the fingerprint, ensuring a more accurate and reliable matching process. Table 1 displays the outcomes obtained by applying the proposed function on different fingerprints.

2.4 *Extracting Local Binary Pattern Vectors*

In this phase, the widely recognized texture descriptor in computer vision, Local Binary Pattern (LBP), is employed. Renowned for its efficacy in delineating image texture, LBP encodes the spatial relationships between the intensity values of the central pixel and its neighboring pixels. The rotation-invariant uniform LBP [8] is applied to fingerprint images in this research to extract feature vectors from dynamic sub-windows (40 $\times$ 40 pixels), with the number of sub-windows adapting to the image size.

The proposed contribution introduced in this phase involves dividing the image into three distinct segments: top, middle, and bottom. This division is essential to ensure that during subsequent translocation mutation steps, the fingerprints are not mixed, preventing the final mutation from placing the bottom sub-window at the top of the image. Consequently, the mutation is based on the sub-window location,

Table 2 Fingerprint division sample

FVC 2004 (102_1)	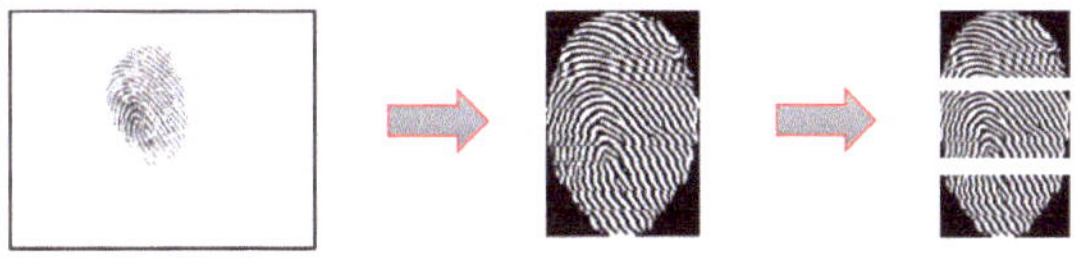

meaning that the mutation for the top part of the image corresponds to the top parts of other images, and likewise for the middle and bottom sections (Table 2).

2.5 *The Modified Bacterial Evolutionary Algorithm for Fingerprint Matching*

At this stage, the algorithm starts by creating clones and performs mutation and gene transfer to produce better clones for matching. In MBAFIS we use permutation encoding where each chromosome is represented by sub-window numbers. Based on experiments, the location of sub-windows matters when processing, hence, the algorithm divides the image into three segments and makes the translocation mutation and gene transfer with respect to the original sub-window location.

The proposed algorithm is explained in the following steps:

1. The algorithm starts by creating eight clones for each person in the database (each person in the database has eight impressions).
2. A fitness function is activated to compute how fit the candidates are to the reference fingerprint and select the best four clones for the next stage.
3. After selecting the best four clones, a translocation mutation is performed with respect to sub-windows locations in the original clones. The result of this step will create two clones.

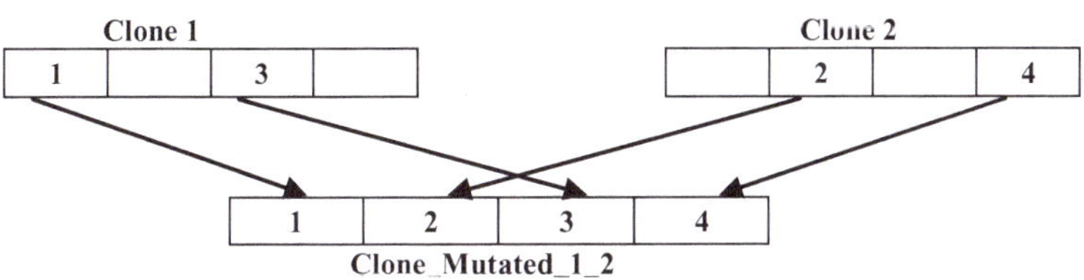

4. The fitness function is performed one more time to determine the best clone among the two (Clone_Mutation_1_2, Clone_Mutation_3_4).
5. Gene transfer (sub-window) is performed between the two clones where the second best is transferring better sub-windows to the best clone so far.

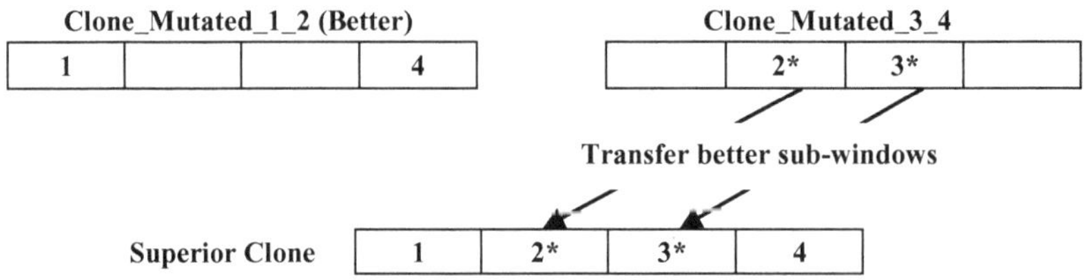

6. The result is one superior clone which will be used to find the similarity score compared to the reference fingerprint using the Euclidian distance.

3 Results and Discussion

The MBAFIS improves identification systems and enhances the ability of forensic experts to identify criminals through fingerprints discovered at crime scenes. Hence experts can support their decision in the final report delivered to the court since unexplainable artificial intelligence algorithms cannot be used.

To assess the system's efficacy, FAR and FRR tests were conducted for the databases FVC 2000 DB1-B, FVC 2000 DB1-B [9], FVC 2002 DB1-B, FVC 2002 DB3-B [10], and FVC 2004 DB1-B [11] yielding the system thresholds. The effectiveness of the system is further illustrated by the Equal Error Rate (EER) values provided in Table 3.

The proposed method was applied to all fingerprints of the mentioned databases. The results of the possible linguistic answers for each database are shown in Table 4.

Table 3 EER values for tested databases

FVC 2000 DB1-B	FVC 2000 DB2-B	FVC 2002 DB1-B	FVC 2002 DB3-B	FVC 2004 DB1-B
0.0	0.0	0.00125	0.0	0.00125

Table 4 List of answers when comparing the fingerprints of the tested databases

	Certainly yes	Very likely	Likely	Possible	Unlikely	Certainly no
FVC 2000 DB1-B	77	3	0	0	0	0
FVC 2000 DB2-B	74	6	0	0	0	0
FVC 2002 DB1-B	72	7	1	0	0	0
FVC 2002 DB3-B	80	0	0	0	0	0
FVC 2004 DB1-B	65	14	1	0	0	0

Table 4 demonstrates the absence of a "Certainly No" response, suggesting a potential connection between the reference fingerprint and the database at some level. The degree of relationship with the database reflects the system's confidence level regarding whether the reference fingerprint belongs to the database. As observed in the table, confidence levels increase as responses approach "Certainly Yes." Conversely, as responses approach "Certainly No," it indicates distortions and significant discrepancies in the fingerprint images within the database.

The proposed system provides a list of three candidates based on the similarity score calculated. Table 5 shows a sample of candidates. Our system has successfully provided the correct answer as the first candidate in all cases, which indicates the system's robustness against all distortions and cuts in the fingerprints of the tested databases.

4 Conclusion

This paper introduces a new fingerprint identification system that presents new methodologies for eliminating undesired portions of fingerprint images, as well as segmenting fingerprint images into three parts (top, middle, bottom) to minimize the number of mutations, thereby enhancing results and processing time. Comprehensive experiments conducted on FVC2000 DB1_B, FVC2000 DB2_B, FVC2002 DB1_B, FVC2002 DB3_B, and FVC 2004 DB1_B databases demonstrate the system's ability to successfully identify whether a fingerprint belongs to the database, along with providing a likelihood measure of its relationship to the database. MBAFIS supports forensics experts with a tool to comprehend relationships between the reference and database fingerprint and facilitates further processing for accurate fingerprint matching.

Table 5 Sample candidates and the correct answers for tested databases (correct matches shown in bold)

Database		Reference fingerprint	Candidate one	Candidate two	Candidate three	Correct answer
FVC 2000 DB1-B	Certainly yes	101_1	**101**	110	107	**101**
		103_5	**103**	108	102	**103**
		108_3	**108**	106	102	**108**
		110_6	**110**	107	103	**110**
	Very likely	102_3	**102**	108	106	**102**
		105_3	**105**	104	106	**105**
		107_1	**107**	110	104	**107**
FVC 2000 DB2-B	Certainly yes	101_6	**101**	107	110	**101**
		103_8	**103**	110	108	**103**
		109_6	**109**	106	102	**109**
		110_3	**110**	103	104	**110**
	Very likely	102_8	**102**	106	108	**102**
		105_1	**105**	104	106	**105**
		106_5	**106**	108	103	**106**
		108_4	**108**	107	102	**108**
FVC 2002 DB1-B	Certainly yes	103_8	**103**	108	104	**103**
		104_1	**104**	108	109	**104**
		105_2	**105**	107	102	**105**
		108_5	**108**	104	102	**108**
		110_8	**110**	108	106	**110**
	Very likely	101_8	**101**	102	103	**101**
		102_4	**102**	108	103	**102**
		106_1	**106**	108	110	**106**
		109_6	**109**	108	103	**109**
	Likely	107_6	**107**	105	108	**107**
FVC 2002 DB3-B	Certainly yes	101_2	**101**	109	104	**101**
		103_6	**103**	104	106	**103**
		105_4	**105**	104	106	**105**
		107_2	**107**	106	104	**107**
FVC 2004 DB1-B	Certainly yes	102_2	**102**	103	109	**102**
		104_6	**104**	103	109	**104**
		106_8	**106**	105	104	**106**
		107_5	**107**	104	108	**107**
	Very likely	102_1	**102**	103	104	**102**
		103_3	**103**	104	102	**103**
		105_2	**105**	107	106	**105**

(continued)

Table 5 (continued)

Database		Reference fingerprint	Candidate one	Candidate two	Candidate three	Correct answer
		109_1	**109**	110	104	**109**
	Likely	101_1	**101**	103	109	**101**

References

1. Monrose, F., Rubin, A.D.: Keystroke dynamics as a biometric for authentication. Futur. Gener. Comput. Syst. **16**(4), 351–359 (2000). https://doi.org/10.1016/S0167-739X(99)00059-X
2. Jayaram, M.A., Fleyeh, H.: Soft computing in biometrics: a pragmatic appraisal. Am. J. Intell. Syst. **2013**(3), 105–112 (2013). https://doi.org/10.5923/j.ajis.20130303.01
3. Momani, A.A., Kóczy, L.T.: A novel fingerprint identification fuzzy system using a center-distance weighted local binary pattern. In: Computational Intelligence and Mathematics for Tackling Complex Problems, vol. 5, p. 1127 (2024). https://doi.org/10.1007/978-3-031-46979-4
4. Lambora, A., Gupta, K., Chopra, K.: Genetic algorithm—a literature review. In: 2019 International Conference on Machine Learning, Big Data, Cloud and Parallel Computing (COMITCon), Faridabad, India, pp. 380–384. IEEE (2019). https://doi.org/10.1109/COMITCon.2019.8862255
5. Kóczy, L.T., Földesi, P., Tüű-Szabó, B.: An effective discrete bacterial memetic evolutionary algorithm for the traveling salesman problem. Int. J. Intell. Syst. **32**(8), 862–876 (2017). https://doi.org/10.1002/int.21893
6. Momani, A.A., Kóczy, L.T.: A robust fingerprint identification fuzzy system using a new rotation method and adaptive core-distance weighted local binary pattern. Pattern Recogn. 1–20 (under review)
7. Chikkerur, S., Wu, C., Govindaraju, V.: LNCS 3072—a systematic approach for feature extraction in fingerprint images. Lecture Notes in Computer Science, vol. 3072. Springer, Berlin, Heidelberg (2007). https://doi.org/10.1007/978-3-540-25948-0_48
8. Ojala, T., Pietikäinen, M., Mäenpää, T.: Multiresolution gray scale and rotation invariant texture classification with local binary patterns. IEEE Trans. Pattern Anal. Mach. Intell. **24**(7), 971–987 (2002). https://doi.org/10.1109/TPAMI.2002.1017623
9. Maio, D., Maltoni, D., Cappelli, R., Wayman, J.L., Jain, A.K.: FVC2000: fingerprint verification competition. IEEE Trans. Pattern Anal. Mach. Intell. **24**(3), 402–412 (2000). https://doi.org/10.1109/34.990140
10. Maio', D., Maltonil, D., Cappellil, R., Wayman', J.L., Jain, A.K.: FVC2002: second fingerprint verification competition. In: 2002 International Conference on Pattern Recognition, Quebec City, QC, Canada, pp. 811–814. IEEE (2002). https://doi.org/10.1109/ICPR.2002.1048144
11. Maio, D., Maltoni, D., Cappelli, R., Wayman, J.L., Jain, A.K.: FVC2004: Third Fingerprint Verification Competition (2004) [Online]. Available at: http://bias.csr.unibo.it/fvc2000

Comparison of Text Similarity Techniques for Power of Attorney Clauses for Polish Banks

Karolina Wadowska and Piotr A. Kowalski

Abstract Text similarity techniques allow for close, but not exactly, matching strings to be compared and extracted from bodies of text. This functionality can be considered very useful in automated processing of the documents. In this paper existing algorithms such as Cosine similarity, Levenhstein distance, pre-trained models etc. are compared and summarised. Based on the attorney clauses from the banking sector—the official formats and given template—we consider the effectuality of each of them. An algorithm selection is made not only based on the similarity score, but also the simplicity of the given solution, often considered an advantage in a highly regulated industry. Nevertheless, this study demonstrated that pre-trained models, in certain instances, exhibit a performance that is twice as effective as other techniques that are more readily explicable in a business context.

Keywords Machine learning · Natural language processing · Pre-trained language models · String matching · Levenshtein distance · Cosine similarity

1 Introduction

Ensuring ongoing customer satisfaction holds significant importance for every business. To maintain this satisfaction, it is crucial to acquire relevant customer data and store it appropriately, such as in databases. This matter, influenced by a relatively high level of regulation, is especially pertinent to the banking sector. The banking and financial sector is progressing at a faster pace compared to other industries, with arti-

K. Wadowska (✉)
AGH Doctoral School, AGH University of Krakow, Krakow, Poland
e-mail: wadowska@agh.edu.pl

P. A. Kowalski
Faculty of Physics and Applied Computer Science, AGH University of Krakow, Kraków, Poland
e-mail: pkowal@agh.edu.pl

Systems Research Institute, Polish Academy of Sciences, Warsaw, Poland

L. T. Kóczy et al. (eds.), *Computational Intelligence and Mathematics for Tackling Complex Problems 6*, Studies in Computational Intelligence 1222,
https://doi.org/10.1007/978-3-031-97879-1_14

ficial intelligence playing a pivotal role. It provides functionalities like cybersecurity and fraud detection, chatbots, assistance in loan processing and decision-making, and proactively safeguards customer information [1].

This article consists of seven sections: Introduction, State of the article, Overview of power of attorney clauses, Datasets, Methodology, Results and Discussion. First two are necessary to introduce the problem, next two are strictly related with data and data quality, Methodology consists the set of approaches used to solve the problem and finally, Results and Discussion sections represent the outcome of the analysis.

In order to make legally sound decisions regarding customer service, banks are required to securely store pertinent data [2]. A common illustration of this is the verification of consent for communication or power of attorney. Complications arise in situations where knowledge is solely available in paper documents. Appropriate tools are then required not only to recognise the text but also to search it to obtain relevant information.

The purpose of this paper is a comparison between text similarity techniques for textual data in the Polish language selected from bank forms. The primary reason for conducting this research is the acknowledgment that dealing with highly inflected languages introduces an additional complexity to string matching. The novelty of this paper is an attempt to test if approaches that are not based on pre-trained models can still be effective for the Polish language and documents that present certain common features.

2 State of the Art

Natural language processing (NLP) has become one of the fastest growing fields of science over the last few years. As a result of technological accomplishments in this area, it is possible for machines to perform actions such as: interpret, manipulate, and comprehend human language [3].

String matching is a fundamental operation in NLP. It tries to find where one or more strings (also called patterns) pop up in a larger string (the search text) [4]. Searching for patterns can be done according to different principles - exact search, approximate search and subphrase search are possible. Exact string matching is the primary way of comparison. It gives a positive result only if an exact match is found. The example of such solution are: Boyer-Moore algorithm, which was designed by Bob Boyer and J. Strother Moore in 1977 [5], Knuth Morris-Pratt (KMP) algorithm [6], founded in the same year and introduced by Donald Knuth, Vaughan Pratt and James H. Morris and Karp-Rabin algorithm presented in 1987 [7]. There are multiple different problems that can be solved using exact string matching. In the following publication [8] it has been used to find patterns in musical databases. In [9], a framework for searching multimedia contents, such as text, image, and audio was proposed.

On the other side, the approximate string matching consists in finding the strings that are similar to the pattern, rather than identical to it. The most popular algorithms in this group include Cosine similarity algorithm, Hamming distance and Levenshtein distance algorithm, few among them will be considered in this paper. Previous researchers have developed algorithms in different directions in dissimilar fields of various applications to fully utilise approximate string matching algorithms. There have been many approaches to string matching and document similarity regarding documents, such as [10, 11] (both in terms of detecting plagiarism). What seems to be also interesting is that topic of spelling correction has been widely followed [12]. There are multiple other practical tasks related with string matching—from basic procedures in test matching for official documents and detecting sequences of symbols, e.g. web page addresses to collaborative filtering [13] and entity reconciliation [14].

Currently, most publications and scientific articles deal with language processing in English. There are also more and more works in other languages. Recently, also automatic document processing in Polish has been the subject of intensive research in several centres in Poland. There are also few users of the Polish language. Therefore it leads to the situation in which articles treating text processing in Polish are rare. However, there are a few topics that are worth noticing. Many papers relating to text processing have been presented recently, e.g. Data Structures Analysis for Text Processing in the Framework of NLP Classification in Polish [15]. The position reviews existing data structures useful for text processing in a particular classification. What is more, some papers reaches similar areas: Towards facts extraction from a text in the Polish language by Tomasz Boiński from Gdansk University of Technology, Gdansk Poland, and Adam Chojnowski from the Gdansk University of Technology [16].

3 Overview of Power of Attorney Clauses

Text analytics represents a tangible application within the banking sector. Currently, the majority of document archives are digitised, but many banks still have a considerable number of customers whose contracts, including powers of attorney, were originally executed in paper format. Regulatory standards necessitate the digitization of such documents. In Poland, regulations governing the electronic document storage practices of banks primarily stem from legislation on the handling of electronic documents and banking laws, particularly the Act on preventing money laundering and terrorism financing. Additionally, these regulations require banks to retain documentation even after the termination of contracts with clients. The imperative to hold information about retail or corporate clients extends over several years [17].

Another aspect of the need to have these records in electronic form is classic or advanced data analysis, which provides greater knowledge about the customer. Documents that are still stored in paper form must be read into text form using OCR (Optical Character Recognition) tools. Especially considering the multitude of Polish characters they usually have limited accuracy. The estimated percentage of errors is 7–16% [18], particularly in cases of poor print quality or the presence of handwritten elements.

The obligation to maintain banking secrecy, as well as the possibilities and scope of exemption from it, are regulated by Article 104 of the Banking Law Act dated August 29, 1997 (Journal of Laws of 2016, item 1988). Among the various statutory exemptions from the duty of banking secrecy, the legislator has provided the option for a person whose confidential information is concerned to authorise the bank to disclose data covered by banking secrecy to a specified entity (Article 104(3)).

4 Datasets

The analysed dataset consists of the clauses from the official websites of Polish banks—they were downloaded manually. The total number of examples is 5, e.g. mBank, Santander Bank, Alior Bank. The authorization to provide information subject to banking secrecy should among others contain:

- indication of the entity that grants the authorization;
- specification of the entity to which the information is to be made available, in particular by providing their name and surname or name, address of residence or registered office;
- indication of the banking activity related to - information subject to banking secrecy and specification of the scope of data covered by banking secrecy that the bank is authorised to transmit.

Each document is a template that can be used by any clients of the bank with filling with personal details. All of the documents given, during the analysis process, where set up with an example of the Infor.pl, which is a specialised portal addressed mainly to: accountants, human resources, business people, company owners and officials. Example of such clause can be find below:

"Ja niżej podpisany XY, zamieszkały w Warszawie ul. XYZ 1/1, PESEL 00000000 001, legitymujacy sie dowodem osobistym seria ABC numer 990099, działajac na podstawie art. 104 ust. 3 ustawy z dnia 29 sierpnia 1997 r. Prawo bankowe, niniejszym upoważniam Bank ABC S.A....." (Eng: I, the undersigned XY, residing in Warsaw, with ID card series ABC number 990099, acting pursuant to Article 104(3) of the Act of August 29, 1997, Banking Law, hereby authorise Bank ABC S.A....) [19].

It is worth noticing that not every financial institution has its own "standardised form." Even so, customers can submit it in any form that complies with the law. This research was aimed at verifying whether, using the proposed approaches, it is possible to obtain an accessible result that would, in the future, facilitate verification of which of the clients has such an authorization if it has not been registered in electronic form.

Preparing the data for textual analysis follows a standard procedure. Normalisation, the process of standardising textual data, involves actions such as converting the text to lowercase, eliminating punctuation, numbers, and special characters (e.g. Polish diacritic signs), and expanding contractions [20]. Language correction was

Table 1 Length (number of characters) of analysed documents

Source	Length before	Length after
mBank	3392	1204
PKO BP	2361	1995
ING Bank Ślaski	1490	597
Alior	975	514
Santander	2762	887
Infor	689	549

omitted as the data quality was already high. The instinctive approach to address the inflection issue was to lemmatise texts. The purpose of lemmatization in natural language processing is to reduce words to their base or root form, known as the lemma. This method proved effective for languages with limited inflection, such as English (see: second example below). Due to the strong inflection of the Polish language, the application of lemmatisation is likely to yield worse results [21].

- jedzcie (Eng: you eat) $\Rightarrow$ jeść (Eng: eating)
- kolację (Eng: dinner) $\Rightarrow$ kolacja (Eng: dinner)

For all approaches, except those using a pre-trained model, lemmatisation was applied.

Table 1 presents the length of all analysed documents before and after preprocessing.

Differences in the lengths of already processed documents result mainly from scoring resulting from leaving spaces for personal data, empty characters and adding sample data in the files, e.g. PESEL number.

5 Methodology

The approach used in this article are divided into two groups: approximate matching techniques and pre-trained algorithms.

5.1 Approximate Matching Techniques

Approximate matching techniques are techniques used to identify the likelihood that two records are a true match based on whether they agree or disagree on the various identifiers like character overlap, edit distance, and phonetic similarity. It works by reviewing the similarity between two strings of text and produces a similarity score.

Here are instances of algorithms within this domain that are used in this article [22]. They are also know as approximate string matching techniques.

Cosine similarity quantifies the similarity between two vectors by assessing the cosine of the angle formed between them. A cosine of 0 corresponds to a value of 1, while for any other angle, the cosine is less than 1, reaching its lowest value at -1. The cosine of the angle between two vectors is indicative of whether they align in a similar direction [23].

Let $A = (a_1, a_2, \ldots, a_n)$, $B = (b_1, b_2, \ldots, b_n)$ be two n-dimensional integer vectors. We examine the cosine similarity between variables A and B and this evaluation will be discussed below [23]. Cosine similarity is a measure of similarity defined as shown in Eq. (1).

$$cosine\,similarity = \frac{A \cdot B}{\|A\| \|B\|} = \frac{\sum_{i=1}^{n} a_i b_i}{\sqrt{\sum_{i=1}^{n} a_i^2} \sqrt{\sum_{i=1}^{n} b_i^2}} \tag{1}$$

To find the similarity between two texts, each text is represented in the form of a vector. As a result A and B represent compared documents. Each word in the text defines a dimension in Euclidean space and the frequency of each element corresponds to the value in the dimension—a_i and b_i represent a term in each of the documents.

Levenshtein distance (edit distance) is the most common measure of the similarity between two strings, which takes into account three types of operations: insertion, deletion and substitution, which are needed to transform one string into another. The higher the distance, the further the two terms are from being identical [24]. Edit distance examples can be found below.

- Kot (Eng: cat) → Kat (Eng: executioner) (Substitution of ‘o’)—Distance: 1
- Szary (Eng: grey) → Sznury (Eng: cords)(Substitution “a” for “u”, Insertion of’n’)—Distance: 2

Mathematically, the value of the Levenshtein distance between two strings A, B, (of length $|A|$ and $|B|$ respectively) can be seen in equation (2). Tail of analysed string is a string of all except the first character and head is considered as the first character of the string. Example notation of $A[n]$ is used to refer the nth element of the string, counting from 0, which means: $A[0] = A_0 = head(A)$.

$$\text{lev(A,B)} = \begin{cases} |A| & if\ |B| = 0, \\ |B| & if\ |A| = 0, \\ lev(tail(A), tail(B)) & if\ head(A) \\ & = head(B), \\ 1 + \min \begin{cases} \text{lev}(tail(A), B) \\ \text{lev}(A, tail(B)) \\ \text{lev}(tail(A), tail(B)) \end{cases} \quad otherwise \end{cases} \tag{2}$$

Which can also be expressed in pseudocode below.

```
int LevenshteinDistance(char s[1..m], char t[1..n])
   // d is a table with m+1 rows and n+1 columns
   declare int d[0..m, 0..n]

   for i from 0 to m
       d[i, 0] := i
   for j from 0 to n
       d[0, j] := j

   for i from 1 to m
       for j from 1 to n
       {
           if s[i] = t[j] then cost := 0
                          else cost := 1
           d[i, j] := minimum(
                              d[i-1, j] + 1, // deletion
                              d[i, j-1] + 1,// insertion
                              d[i-1, j-1] + cost // substitution
                          )
       }

   return d[m, n]
```

In summary, cosine distance focuses on the angular relationship between vectors, while Levenshtein distance focuses on the amount of editing operations needed to transform one sequence into another.

N-gram is defined as is a sequence of n adjacent symbols in particular order. Levenshshtein operations presented below only included single character operations, so called unigrams. N-gram can be therefore introduced as a set of values generated from a string by pairing sequentially occurring 'n' characters/words. In Table 2, examples for a string "alamakota" (Eng: Ala has a cat) in different n-gram variations has been presented.

Table 2 Sentence "Ala ma kota" on different n-gram levels

Name	n	grams
Unigram	1	a,l,a,m,a,k,o,t,a
Bigram	2	al, la, am, ma, ak, ka, ot, ta
Trigram	3	ala, lam, ama, mak, oko, kot, ota

Let's say we want to determine the probability of the sentence, "Which is the best car insurance package". Based on Unigram language model, probability can be calculated as following:

[$latex$]P("Which is best car insurance package") = [/latex]

The Levenshtein distance is just a special case of the n-gram distance in the case of n = 1 (unigrams) [25]. As mentioned below, the distance is calculated between each character. The n-gram distance model was created to capture multiple characters at the time. The implementation of n-gram in this study is based on calculation Levenshtein distance between 3-gram representations in the text.

5.2 *Pre-train Language Model*

Pre-trained language models are powerful tools for text similarity tasks because they are able to work with both: semantic and syntactic information. The concept itself is related to transfer learning. Transfer learning is a machine learning methodology that entails utilising knowledge gained from addressing one task and applying it to accomplish another task, which is relatively similar in the field [26].

BERT (Bidirectional Encoder Representations from Transformers) is a pre-trained language model based on transformers which is extensively used for a range of natural language processing tasks, including the assessment of text similarity [27]. It was introduced in October 2018 by researchers at Google. For pre-train model only basic cleaning of text data was introduced, e.g. removing punctuation. The exact model used in this exercises was named bert-base-multilingual-cased and can be find on Hugging Face open source portal.

6 Results

In the scenario outlined in this article, a comparison of average similarity values can be conducted. Similarity scores of each approach has been calculated according to its specifics, mentioned in Methodology section. 0 is considered to be no similarity and 1 as the maximum similarity. Given that the presented clauses in bank powers of

Table 3 Results of approximate string techniques comparison

Similarity metrics				
Bank name	Approximate matching techniques			Pre-trained algorithms
	Cosine similarity	Levenshtein distance	3-Gram	BERT
mBank	0.36	0.27	0.38	0.89
PKO BP	0.36	0.22	0.29	0.88
ING Bank Ślaski	0.53	0.24	0.35	0.82
Alior	0.37	0.29	0.35	0.84
Santander	0.49	0.28	0.44	0.95
$\overline{R}$	0.42	0.26	0.36	0.88

attorney adhere to statutory requirements, they are theoretically considered to have semantically identical or almost identical content. Therefore, the most effective algorithm would be the one yielding the highest average similarity value for documents. This algorithm evaluates the similarity between two text strings, generating a similarity score based on factors such as character overlap, edit distance, and phonetic similarity. A higher score indicates a greater likelihood that the strings match.

In this publication, text data sets were verified, each of which represented a power of attorney template specific to one of the Polish banks, and string similarity techniques such as: cosine similarity, Levenshtein distance, n-gram and pre-trained language models. The results were presented in Table 3.

The results obtained reveal notable distinctions between the pre-trained model and conventional string matching. Nevertheless, it is evident that none of the methodologies demonstrates markedly poor outcomes.

The last line of Table 3 contains the arithmetic mean of the results obtained for all documents. While pre-trained models undoubtedly provide the best results 0.88, the use of cosine distance comes next 0.42. However, this difference is more than double. The n-gram method using trigrams achieved slightly lower results than cosine distance 0.36. Levenshtein distance comes in last place 0.26.

It should be noted that the difference that emerged between approximate techniques and pre-trained models was quite expected. Levenshtein distance however has no knowledge of semantics, it's simply an edit distance calculated for documents. In practice, Cosine distance is used on an embedding that encodes semantic information: when looking for words or documents (collections of words) that are using similar words in similar contexts. Which is technically applied in presented case. As given below, all of the simpler approaches has not been sufficient for given case. The analysis below was made with the assumption that the difference between a more advanced model, such as BERT, and simpler methods might be less significant. This would mean that there is a real sense in applying this type of methods to the Polish language, assuming that the documents are somewhat schematic.

Ultimately, only the results of pre-trained models could be applied in business terms. As mentioned above, a large part of the documents still exist in institutions in paper form. The purpose of the institution, which is indicated here as the subject of the study, is to determine which document meets the requirements for being a power of attorney. For this purpose, it must be read into an electronic form, which, as mentioned earlier, is done with a certain margin of error. Therefore, technique such as cosine similarity and Levenshtein distance, which achieve lower results, are most likely to have no practical application.

7 Discussion, Contribution, Implication, Limitations and Further Research

It is very difficult to refer in the discussion to the use of similar techniques on similar data in Polish for other articles. While there are several issues related to tagging in the literature, it is difficult to compare the results due to the uniqueness of the business and data. Approaches also seems to be different—in this case, a decision was made to compare quite basic techniques from the area of language analysis with advanced ones, such as pre-trained models. There are no results for this unique dataset using data tagging, classification, or other known approaches.

Although artificial intelligence substantially improves the functioning of the financial sector, it is crucial to recognize that its application entails inherent risks. Therefore, oversight should be applied to actions carried out by artificial intelligence. Additionally, integrating AI in banking introduces potential risks and threats to the security of customer data. This is why the banking industry frequently favors algorithms characterised by a high degree of explainability. Consequently, the topic covers a diverse range of methodologies, spanning from well-established approaches to comparatively recent innovations such as the pre-trained model.

This paper will have the biggest implications for research related to the Polish language and the banking industry. While it is known that embeddings are included in models like BERT, and similar, and try to reflect some level of semantic knowledge into their encoding. That is, words that tend to have similar meanings will end up close together in the high-dimensional embedding space. However, it seemed possible to use simpler approaches, which would make it possible to simplify the issue and quickly implement it in practice.

It can certainly be said that one of the biggest limitations related to this issue is the strong flexibility of the Polish language, which makes some issues impossible to solve with simple approaches. However, it is worth noting that the applications of AI for the banking sector are very specific—most often they require the use of only algorithms with a high level of explainability.

As part of the development of the above research, it is certainly worth taking a look at the appropriate data preparation process. Polish is a highly inflected language and not all tools have the same accuracy for it as for e.g. English. The next steps may be research towards modifying metrics already analysed or others not included in the study, such as: Jaro-Winkler distance and Jaccard similarity coefficient.

Acknowledgements This work was partially supported by the program "Excellence initiative—research university" for the AGH University of Krakow and by Grant for Statutory Activity from the Faculty of Physics and Applied Computer Science of the AGH University of Krakow.

References

1. Rana, A., Bisht, D., Pandey, S., Singh, R., Chhabra, G., Joshi, K.: Artificial intelligence indulgence in banking and financial sectors, pp. 1–5 (2023). https://doi.org/10.1109/InC457730.2023.10263088
2. Ustawa z dnia 29 sierpnia 1997 r. prawo bankowe (1997), z późniejszymi zmianami
3. Chowdhury, G.: Natural language processing. Annu. Rev. Inf. Sci. Technol. **37**(1), 51–89 (2003). https://doi.org/10.1002/aris.1440370103
4. Bhagya Sri, M., Bhavsar, R., Narooka, P.: String matching algorithms. Int. J. Eng. Comput. Sci. **7**(03), 23769–23772 (2018)
5. Boyer, R.S., Moore, J.S.: A fast string searching algorithm. Commun. ACM **20**(10), 762–772 (1977)
6. Knuth, D.E., Morris, J.H., Jr., Pratt, V.R.: Fast pattern matching in strings. SIAM J. Comput. **6**(2), 323–350 (1977)
7. Karp, R.M., Rabin, M.O.: Efficient randomized pattern-matching algorithms. IBM J. Res. Dev. **31**(2), 249–260 (1987). https://doi.org/10.1147/rd.312.0249
8. Szeto, W.M., Wong, M.H.: Stream segregation algorithm for pattern matching in polyphonic music databases. Multimedia Tools Appl. **30**, 109–127 (2006). https://doi.org/10.1007/s11042-006-0011-9
9. Shrivastav, S., Kumar, S., Kumar, K.: Towards an ontology based framework for searching multimedia contents on the web. Multimedia Tools Appl. **76**, 1–30 (2017). https://doi.org/10.1007/s11042-017-4350-5
10. Rajkumar Kundu, K.K.: Contextual plagiarism detection using latent semantic analysis. Int. Res. J. Adv. Eng. Sci. **2**, 214–217 (2017)
11. Gipp, B., Meuschke, N., Beel, J.: Comparative evaluation of text- and citation-based plagiarism detection approaches using GuttenPlag. In: Proceedings of 11th Annual International ACM/IEEE-CS Joint Conference on Digital Libraries (JCDL'11). ACM, Ottawa, Canada (2011). https://doi.org/10.1145/1998076.1998124
12. Mawardi, V.C., Rudy, R., Naga, D.S.: Fast and accurate spelling correction using Trie and Damerau-Levenshtein distance bigram. TELKOMNIKA: Indonesian J. Electr. Eng. **16**, 827–833 (2018)
13. Bayardo, R.J., Ma, Y., Srikant, R.: Scaling up all pairs similarity search. In: Proceedings of the 16th International Conference on World Wide Web, pp. 131–140. Association for Computing Machinery, New York, NY, USA (2007). https://doi.org/10.1145/1242572.1242591
14. Dey, D., Sarkar, S., De, P.: A distance-based approach to entity reconciliation in heterogeneous databases. IEEE Trans. Knowl. Data Eng. **14**, 567–582 (2002)
15. Krzeszewska, U., Poniszewska-Marańda, A.: Data structures analysis for text processing in the framework of nlp classification in polish. In: 2022 International Conference on Software, Telecommunications and Computer Networks (SoftCOM), pp. 1–6 (2022). https://doi.org/10.23919/SoftCOM55329.2022.9911412
16. Boiński, T., Chojnowski, A.: Towards facts extraction from text in Polish language. In: 2017 IEEE International Conference on INnovations in Intelligent SysTems and Applications (INISTA), pp. 13–17 (2017). https://doi.org/10.1109/INISTA.2017.8001124

17. Ustawa z dnia 1 marca 2018 r. o przeciwdziałaniu praniu pieniędzy oraz finansowaniu terroryzmu (2018), przeciwdziałanie praniu pieniędzy oraz finansowaniu terroryzmu
18. Navarro, G.: A guided tour to approximate string matching. ACM Comput. Surv. **33**(1), 31–88 (2001). https://doi.org/10.1145/375360.375365
19. Infor website. https://www.infor.pl/prawo/umowy/finanse/751744,Upowaznienie-do-przekazania-danych-objetych-tajemnica-bankowa.html. Accessed 10 Dec 2023
20. Qaiser, S., Ali, R.: Text mining: use of TF-IDF to examine the relevance of words to documents. Int. J. Comput. Appl. **181** (2018). https://doi.org/10.5120/ijca2018917395
21. Piskorski, J., Sydow, M.: Usability of string distance metrics for name matching tasks in Polish. Wydawnictwo Poznanskie Sp. z o. o., Poznan (Poland) (2007). http://www.ltc.amu.edu.pl/content.en.html
22. Charalampopoulos, P., Kociumaka, T., Wellnitz, P.: Faster approximate pattern matching: a unified approach (2020)
23. Peng, J.: Research on the text similarity algorithms in automatic scoring of subjective questions. J. Phys.: Conf. Ser. **1952**(4), 042039 (2021). https://doi.org/10.1088/1742-6596/1952/4/042039
24. Levenshtein, V.I.: Binary codes capable of correcting deletions, insertions and reversals. Soviet Physics Doklady **10**(8), 707–710 (1966); doklady Akademii Nauk SSSR 163(4):845–848 (1965)
25. Jurafsky, D., Martin, J.: Speech and Language Processing: An Introduction to Natural Language Processing, Computational Linguistics, and Speech Recognition, vol. 2. Prentice Hall (2008)
26. Bahl, L., Jelinek, F., Mercer, R.: A maximum likelihood approach to continuous speech recognition. IEEE Trans. Pattern Anal. Mach. Intell. **PAMI-5**, 179–190 (1983). https://doi.org/10.1109/TPAMI.1983.4767370
27. Devlin, J., Chang, M.W., Lee, K., Toutanova, K.: BERT: pre-training of deep bidirectional transformers for language understanding. In: Burstein, J., Doran, C., Solorio, T. (eds.) Proceedings of the 2019 Conference of the North American Chapter of the Association for Computational Linguistics: Human Language Technologies, vol. 1 (Long and Short Papers), pp. 4171–4186. Association for Computational Linguistics, Minneapolis, Minnesota (2019). https://doi.org/10.18653/v1/N19-1423, https://aclanthology.org/N19-1423

Belief Change: Axiomatic Characterization of KM-Erasure

Eduardo Fermé

Abstract One of the most important models in the literature of belief change is *update*, defined by Katsuno and Mendelzon in 1992. In their work, KM mentioned *erasure* as the counterpart change of update. However, erasure was only defined at the basic level. In this paper: (1) we complete the axiomatics of erasure and provide its relation with update via the Levi and Harper Identities, and (2) we provide a semantics in terms of possible worlds.

Keywords Belief change · Katsuno and Mendelzon update · Erasure · Axiomatic characterization · Possible world semantics

1 Introduction

The logic of theory change emerged as a significant focus in philosophical logic and artificial intelligence in the mid-1980s. The influential AGM model of belief change, proposed by Alchourrón, Gärdenfors, and Makinson in [1], serves as a formal framework for characterizing the dynamics and belief state of a rational agent.

Within the AGM framework, beliefs are ideally represented by sets of beliefs, which are deductively closed sets of sentences. A change involves the addition or removal of a specific sentence from a set of beliefs to obtain a new set of beliefs. AGM recognizes three types of changes: **expansion**, which involves incorporating sentences into the original belief set without eliminating any sentences from it; **revision**, which incorporates a sentence into the original belief set but may eliminate some beliefs to maintain consistency in the revised set; and **contraction**, which eliminates sentences from the original set without incorporating any new ones. Revision is the counterpart of contraction, and they can be mutually defined through the Levi and Harper identities.

E. Fermé (✉)
Universidade da Madeira, Funchal, Portugal
e-mail: ferme@uma.pt

NOVA-LINCS, Caparica, Portugal

L. T. Kóczy et al. (eds.), *Computational Intelligence and Mathematics for Tackling Complex Problems 6*, Studies in Computational Intelligence 1222,
https://doi.org/10.1007/978-3-031-97879-1_15

In 1992, Katsuno and Mendelzon introduced a new belief change operator called **update** [7]. While AGM operators are well-suited for capturing changes reflecting evolving knowledge about a static world, KM-update operators are designed to represent changes in an agent's beliefs resulting from changes in the world. This distinction was initially highlighted by Keller and Winslett in [8] within the context of relational databases and is exemplified in the subsequent situation [13]:

Initially the agent knows that there is either a book on the table (α) or a magazine on the table (β), but not both.

Case 1: The agent is told that there is a book on the table. She concludes that there is no magazine on the table. This is revision.
Case 2: The agent is told that subsequently a book has been put on the table. In this case she should not conclude that there is no magazine on the table. This is update.

Just as belief revision has its counterpart in contraction, KM proposed erasure as the counterpart to update. "Intuitively, erasing μ means the world may have changed in such a way that μ is not true. In contrast, contracting μ means our description of the set of possible worlds must be adjusted to the possibility of μ being false" [7, p. 194]. However, they did not fully characterize it, providing only a basic and finite set-level description.

In this paper, we present a comprehensive axiomatic characterization of erasure for belief sets at both the basic and supplementary levels. We demonstrate its relationship with update through Levi and Harper identities. Finally, we provide a semantic characterization.

The paper is organized as follows: In Sect. 2, we introduce the necessary notations and background concepts. In Sect. 3, we present and fully characterize the model of erasure, establishing the connection of this axiomatization with update. In Sect. 4, we introduce semantics for erasure operators. Section 5 is devoted to conclusions and future works. Lastly, in the Appendix, we provide sketch proofs for all original results presented in the paper.

2 Background

2.1 Formal Preliminaries

We will assume a propositional language $\mathcal{L}$ that contains the usual truth functional connectives: $\neg$ (negation), $\wedge$ (conjunction), $\vee$ (disjunction), $\rightarrow$ (implication) and $\leftrightarrow$ (equivalence). We shall make use of a consequence operation Cn that takes sets of sentences to sets of sentences and which satisfies the standard Tarskian properties, namely: (i) $A \subseteq Cn(A)$ (*inclusion*); (ii) If $A \subseteq B$, then $Cn(A) \subseteq Cn(B)$ (*monotony*) and (iii) $Cn(A) = Cn(Cn(A))$ (*iteration*). Furthermore we will assume that Cn satisfies the following three properties: (iv) If α can be derived from A by classical truth-functional logic, then $\alpha \in Cn(A)$ (*supraclassicality*); (v) $\beta \in Cn(A \cup \{\alpha\})$ if and only if $\alpha \to \beta \in Cn(A)$ (*deduction*) and (vi) If $\alpha \in Cn(A)$, then $\alpha \in Cn(A')$ for some finite subset A' of A (*compactness*). We will sometimes use $Cn(\alpha)$ for $Cn(\{\alpha\})$, $A \vdash \alpha$ for $\alpha \in Cn(A)$, $\alpha \vdash \beta$ for $\{\alpha\} \vdash \beta$, $\vdash \alpha$ for $\alpha \in Cn(\emptyset)$, $A \nvdash \alpha$ for $\alpha \notin Cn(A)$, $\alpha \nvdash \beta$ for $\{\alpha\} \nvdash \beta$, $\nvdash \alpha$ for $\alpha \notin Cn(\emptyset)$. Greek letters $\alpha, \beta, \ldots$ will be used to denote sentences of $\mathcal{L}$. H, K are reserved to represent a set of sentences that are closed under logical consequence (i.e. $K = Cn(K)$)—such a set is called a *belief set*. $K_\perp$ is the inconsistent belief set. A belief set is *complete* if and only if for all $\alpha \in \mathcal{L}$, $\alpha \in K$ or $\neg\alpha \in K$. The expansion (denoted by $+$) of a belief set by a sentence is a two steps procedure: first that sentence is added to the belief set and, afterwards, the resulting set is closed by logical consequence; *i.e.*, given a belief set K, the outcome of the expansion of K by α, which is denoted by $K + \alpha$, is the set $Cn(K \cup \{\alpha\})$.

A possible world is a maximal consistent subset of $\mathcal{L}$. The set of all possible worlds will be denoted by Ω. The set of possible worlds that contain $R \subseteq \mathcal{L}$ is denoted by $[[R]]$, i.e., $[[R]] = \{M \in \Omega : R \subseteq M\}$. If R is inconsistent, then $[[R]] = \emptyset$. The elements of R are designated by $R-$worlds. For any sentence α, $[[\alpha]]$ is an abbreviation of $[[Cn(\{\alpha\}]]$ and its elements are designated by α-worlds. For all $V \subseteq \Omega$, we define the associate theory of V as $Th(V) = \bigcap\{\omega \in \Omega : \omega \in V\}$.

Given a set of worlds M, we denote by α_M a formula such that $[[\alpha_M]] = M$. We often omit the braces, by writing, e.g., $\alpha_{\omega,\omega'}$ instead of $\alpha_{\{\omega,\omega'\}}$. We will also note $K_M = Cn(\alpha_M)$. If $\leq$ is a preorder (a reflexive and transitive relation), then $\simeq$ is a notation for the associated equivalence relation ($a \simeq b$ iff $a \leq b$ and $b \leq a$), and $<$ is the notation for the associated strict order ($a < b$ iff $a \leq b$ and $b \nleq a$). We denote by $\min(M, \leq)$ the set $\min(M, \leq) = \{\omega \in M \mid \forall \omega' \in M, \omega' \nless \omega\}$.

2.2 KM-Update

As mentioned in the introduction, Katsuno and Mendelzon proposed a model to depict the alteration of beliefs resulting from modifications in the objects of belief. An update operator $\diamond$ [7] is a mapping associating a belief set K (the agent's belief set) and α (the incoming information/piece of evidence) with a new belief set $K \diamond \alpha$ (the updated belief set). The postulates for update are as follows[1]:

Definition 1 (*KM update operator* [7]) An operator $\diamond$ is a *KM update operator* if for all belief sets K, H and all formulae α, β, the following conditions are satisfied:

(K$\diamond$0) $K \diamond \alpha$ is a belief set whenever K is a belief set.
(K$\diamond$1) $\alpha \in K \diamond \alpha$.
(K$\diamond$2) If $\alpha \in K$ then $K \diamond \alpha = K$.
(K$\diamond$3) If $\nvdash \neg\alpha$ and $K \neq K_\perp$, then $K \diamond \alpha \neq K_\perp$.

[1] In the original KM formulation, K must be finitely axiomatizable. To overcome this restriction, we use the rewritten version of the postulates provided by Peppas and Williams in [11].

(K◇4) If $\vdash \alpha \leftrightarrow \beta$ then $K \diamond \alpha = K \diamond \beta$.
(K◇5) $K \diamond (\alpha \wedge \beta) \subseteq (K \diamond \alpha) + \beta$
(K◇6) If $\beta \in K \diamond \alpha$ and $\alpha \in K \diamond \beta$, then $K \diamond \alpha = K \diamond \beta$.
(K◇7) If K is complete, $K \diamond (\alpha \vee \beta) \subseteq Cn(K \diamond \alpha \cup K \diamond \beta)$
(K◇8) $K \diamond \alpha = \bigcap_{K \subseteq H} H \diamond \alpha$

In their seminal paper, Katsuno and Mendelzon provided a semantic characterization in terms of possible worlds. For a detailed explanation of each postulate, refer to [6]. Additional studies on belief update can be found in [2–4, 9, 12].

3 Erasure

As previously mentioned, Erasure was proposed by Katsuno and Mendelzon as a counterpart of update, where erasing a belief α means the world may have changed in such a way that α is not true. In [7], they proposed the following postulates for Erasure:

(K$\mathbin{-\!\!\diamond\!\!-}$0) $K \mathbin{-\!\!\diamond\!\!-} \alpha$ is a belief set whenever K is a belief set.
(K$\mathbin{-\!\!\diamond\!\!-}$1) $K \mathbin{-\!\!\diamond\!\!-} \alpha \subseteq K$
(K$\mathbin{-\!\!\diamond\!\!-}$2) If $K \vdash \neg\alpha$ then $K \mathbin{-\!\!\diamond\!\!-} \alpha = K$.
(K$\mathbin{-\!\!\diamond\!\!-}$3) If $K \neq K_{\perp}$ and $\nvdash \alpha$, then $K \mathbin{-\!\!\diamond\!\!-} \alpha \nvdash \alpha$.
(K$\mathbin{-\!\!\diamond\!\!-}$4) If $\vdash \alpha \leftrightarrow \beta$ then $K \mathbin{-\!\!\diamond\!\!-} \alpha = K \mathbin{-\!\!\diamond\!\!-} \beta$.
(K$\mathbin{-\!\!\diamond\!\!-}$R) $K \subseteq (K \mathbin{-\!\!\diamond\!\!-} \alpha) + \alpha$.
(K$\mathbin{-\!\!\diamond\!\!-}$8) $K \mathbin{-\!\!\diamond\!\!-} \alpha = \bigcap_{K \subseteq H} H \mathbin{-\!\!\diamond\!\!-} \alpha$

These postulates only define erasure for a single belief. To extend erasure to the supplementary level, i.e., erasing a conjunction of beliefs, additional postulates are required (similar to (K◇5)–(K◇7) in update):

Definition 2 (*Erasure operator*) An operator $\mathbin{-\!\!\diamond\!\!-}$ is a *erasure operator* if and only if for all belief sets K, H and all formulae α, β $\mathbin{-\!\!\diamond\!\!-}$ satisfies (K$\mathbin{-\!\!\diamond\!\!-}$0) - (K$\mathbin{-\!\!\diamond\!\!-}$4), (K$\mathbin{-\!\!\diamond\!\!-}$R), (K$\mathbin{-\!\!\diamond\!\!-}$8) and

(K$\mathbin{-\!\!\diamond\!\!-}$5) $(K \mathbin{-\!\!\diamond\!\!-} \alpha) \cap (K \mathbin{-\!\!\diamond\!\!-} \beta) \subseteq K \mathbin{-\!\!\diamond\!\!-} (\alpha \wedge \beta)$
(K$\mathbin{-\!\!\diamond\!\!-}$6) If $\beta \rightarrow \alpha \in K \mathbin{-\!\!\diamond\!\!-} \alpha$ and $\alpha \rightarrow \beta \in K \mathbin{-\!\!\diamond\!\!-} \beta$, then $K \mathbin{-\!\!\diamond\!\!-} \alpha = K \mathbin{-\!\!\diamond\!\!-} \beta$.
(K$\mathbin{-\!\!\diamond\!\!-}$7) If K is complete, $K \mathbin{-\!\!\diamond\!\!-} (\alpha \wedge \beta) \subseteq Cn((K \mathbin{-\!\!\diamond\!\!-} \alpha) \cup (K \mathbin{-\!\!\diamond\!\!-} \beta))$.

3.1 Relation Between Erasure and Update

We have observed that erasure and update are characterized by two distinct sets of postulates. These postulates are independent in the sense that the postulates of update do not refer to erasure, and vice versa. However, it is possible to define update functions in terms of erasure functions, and vice versa. We can define update in terms of erasure through the Levi identity [10]:

$K \diamond \alpha = (K \mathrel{-\!\!\diamond\!\!-} \neg\alpha) + \alpha$ (Levi identity)

Theorem 1 *Let $\mathrel{-\!\!\diamond\!\!-}$ be an operator defined as in Definition 2. Let $\diamond$ be defined from $\mathrel{-\!\!\diamond\!\!-}$ via the Levi identity. Then:*

1. *$\diamond$ satisfies the update postulates ($K\diamond0$) to ($K\diamond4$).*
2. *If $\mathrel{-\!\!\diamond\!\!-}$ also satisfies ($K\mathrel{-\!\!\diamond\!\!-}5$) and ($K\mathrel{-\!\!\diamond\!\!-}R$), then $\diamond$ satisfies ($K\diamond5$).*
3. *If $\mathrel{-\!\!\diamond\!\!-}$ also satisfies ($K\mathrel{-\!\!\diamond\!\!-}6$), then $\diamond$ satisfies ($K\diamond6$).*
4. *If $\mathrel{-\!\!\diamond\!\!-}$ also satisfies ($K\mathrel{-\!\!\diamond\!\!-}7$), then $\diamond$ satisfies ($K\diamond7$).*
5. *If $\mathrel{-\!\!\diamond\!\!-}$ also satisfies ($K\mathrel{-\!\!\diamond\!\!-}8$), then $\diamond$ satisfies ($K\diamond8$).*

To define erasure in terms of update we use the following identity [5]:

$$(K \mathrel{-\!\!\diamond\!\!-} \alpha) = (K \diamond \neg\alpha) \cap K \text{ (Harper identity)}$$

Similar relationships between the postulates hold in this direction as well.

Theorem 2 *Let $\diamond$ be an operator defined as in Definition 1. Let $\mathrel{-\!\!\diamond\!\!-}$ be defined from $\diamond$ via the Harper identity. Then:*

1. *$\mathrel{-\!\!\diamond\!\!-}$ satisfies the erasure postulates ($K\mathrel{-\!\!\diamond\!\!-}0$) to ($K\mathrel{-\!\!\diamond\!\!-}4$) and ($K\mathrel{-\!\!\diamond\!\!-}R$).*
2. *If $\diamond$ also satisfies ($K\diamond5$), then $\mathrel{-\!\!\diamond\!\!-}$ satisfies ($K\mathrel{-\!\!\diamond\!\!-}5$).*
3. *If $\diamond$ also satisfies ($K\diamond6$), then $\mathrel{-\!\!\diamond\!\!-}$ satisfies ($K\mathrel{-\!\!\diamond\!\!-}6$).*
4. *If $\diamond$ also satisfies ($K\diamond7$), then $\mathrel{-\!\!\diamond\!\!-}$ satisfies ($K\mathrel{-\!\!\diamond\!\!-}7$).*
5. *If $\diamond$ also satisfies ($K\diamond8$), then $\mathrel{-\!\!\diamond\!\!-}$ satisfies ($K\mathrel{-\!\!\diamond\!\!-}8$).*

4 Semantic Characterization of Erasure

In this section, we will semantically characterize erasure operators in terms of possible worlds through a faithful assignment, as defined in [7] for the characterization of update:

Definition 3 (*Faithful assignment* [7]) A *faithful assignment* is a mapping associating each world $\omega_i \in \Omega$ with a partial preorder $\leq_{\omega_i}$ over Ω such that for all $\omega_i, \omega \in \Omega$, if $\omega \neq \omega_i$ then $\omega_i <_{\omega_i} \omega$.

Through a faithful assignment, we can establish a semantic characterization of erasure:

Theorem 3 *An operator $\mathrel{-\!\!\diamond\!\!-}$ is a KM erasure operator if and only if there exists a faithful assignment $\omega_i \mapsto \leq_{\omega_i}$ such that for all belief set K and all formula α,*

$$[[K \mathrel{-\!\!\diamond\!\!-} \alpha]] = \left(\bigcup_{\omega_i \in [[K]]} \min([[\neg\alpha]], \leq_{\omega_i}) \right) \cup [[K]].$$

5 Conclusions and Future Works

In this paper, we complete the axiomatization of erasure, a contraction function that reflects changes in the world where no new information is added to the current state of belief. In future works, we will explore constructive methods for erasure operators.

Acknowledgements This work was partially supported by FCT-Fundação para a Ciência e a Tecnologia, Portugal through projects UIDB/04516/2020 (NOVA LINCS) and PTDC/CCICOM/4464/2020.

Appendix: Proofs

Proof of Theorem 1 [2] Let K be a theory and $-\!\!\diamond\!\!-$ an operation for K that satisfies the erasure postulates (K$-\!\!\diamond\!\!-$0) to (K$-\!\!\diamond\!\!-$4) and let $\diamond$ be defined from $-\!\!\diamond\!\!-$ via the Levi identity. (K$\diamond$0), (K$\diamond$1), (K$\diamond$2), (K$\diamond$3), (K$\diamond$4) and (K$\diamond$8) follow trivially. For (K$\diamond$5), let $\epsilon \in K \diamond (\alpha \wedge \beta)$. By Levi identity $\epsilon \in (K -\!\!\diamond\!\!- \neg(\alpha \wedge \beta)) + (\alpha \wedge \beta)$. We need to prove that $\epsilon \in (K \diamond \alpha) + \beta$, which is equivalent to Levi identity to prove that $\epsilon \in (K -\!\!\diamond\!\!- \neg\alpha) + \alpha + \beta = (K -\!\!\diamond\!\!- \neg\alpha) + (\alpha \wedge \beta)$. Since $\neg\alpha \equiv \neg(\alpha \wedge \beta) \wedge (\neg\alpha \vee \beta)$, then by (K$-\!\!\diamond\!\!-$5) it is suffice to prove that (a) $\epsilon \in (K -\!\!\diamond\!\!- \neg(\alpha \wedge \beta)) + (\alpha \wedge \beta)$ and (b) $\epsilon \in (K -\!\!\diamond\!\!- (\neg\alpha \vee \beta)) + (\alpha \wedge \beta)$. (a) follows by the initial hypothesis. It follows from (a) and deduction that $(\alpha \wedge \beta) \rightarrow \epsilon \in K -\!\!\diamond\!\!- \neg(\alpha \wedge \beta)$, from which it follows by (K$-\!\!\diamond\!\!-$1) that $(\alpha \wedge \beta) \rightarrow \epsilon \in K$. By (K$-\!\!\diamond\!\!-$R) we obtain that $K \subseteq K \in (K -\!\!\diamond\!\!- (\neg\alpha \vee \beta) + (\neg\alpha \vee \beta)$ and $(K -\!\!\diamond\!\!- (\neg\alpha \vee \beta) + (\neg\alpha \vee \beta) \subseteq (K -\!\!\diamond\!\!- (\neg\alpha \vee \beta)) + (\alpha \wedge \beta)$, from which it follows that $(\alpha \wedge \beta) \rightarrow \epsilon \in (K -\!\!\diamond\!\!- (\neg\alpha \vee \beta)) + (\alpha \wedge \beta)$, then $\epsilon \in (K -\!\!\diamond\!\!- (\neg\alpha \vee \beta)) + (\alpha \wedge \beta)$. Hence $\epsilon \in (K \diamond \alpha) + \beta$. For (K$\diamond$6), let $\beta \in K \diamond \alpha$ and $\alpha \in K \diamond \beta$. If $K = K_{\perp}$, then it follows by previous (K$\diamond$2) that $K \diamond \alpha = K \diamond \beta = K$. If $\vdash \alpha$, then by (K$\diamond$2) $K \diamond \alpha = K$ from which it follows that $\beta \in K$ and again by (K$\diamond$2) $K \diamond \beta = K$. Idem if $\vdash \beta$. Let $K \neq K_{\perp}, \nvdash \alpha$ and $\nvdash \beta$. By Levi identity, $\beta \in (K -\!\!\diamond\!\!- \neg\alpha) + \alpha$ and $\alpha \in (K -\!\!\diamond\!\!- \neg\beta) + \beta$. Then by deduction $(\neg\alpha \vee \beta) \in (K -\!\!\diamond\!\!- \neg\alpha)$ and $(\neg\beta \vee \alpha) \in (K -\!\!\diamond\!\!- \neg\beta)$. (K$-\!\!\diamond\!\!-$0) yields that $\neg\beta \rightarrow \neg\alpha \in (K -\!\!\diamond\!\!- \neg\alpha)$ and $\neg\alpha \rightarrow \neg\beta \in (K -\!\!\diamond\!\!- \neg\beta)$, from which it follows by (K$-\!\!\diamond\!\!-$6) that $(K -\!\!\diamond\!\!- \neg\alpha) = (K -\!\!\diamond\!\!- \neg\beta)$. We will prove by double inclusion that $(K -\!\!\diamond\!\!- \neg\alpha) + \alpha = (K -\!\!\diamond\!\!- \neg\beta) + \beta$. Let $\epsilon \in (K -\!\!\diamond\!\!- \neg\alpha) + \alpha$. Then $\alpha \rightarrow \epsilon \in (K -\!\!\diamond\!\!- \neg\alpha)$, from which it follows that $\alpha \rightarrow \epsilon \in (K -\!\!\diamond\!\!- \neg\beta)$ and since $\alpha \in (K -\!\!\diamond\!\!- \neg\beta) + \beta$, we obtain $\epsilon \in (K -\!\!\diamond\!\!- \neg\beta) + \beta$, from which it follows that $(K -\!\!\diamond\!\!- \neg\alpha) + \alpha \subseteq (K -\!\!\diamond\!\!- \neg\beta) + \beta$. Due to the symmetry of the case, $(K -\!\!\diamond\!\!- \neg\beta) + \beta \subseteq (K -\!\!\diamond\!\!- \neg\alpha) + \alpha$, that concludes the proof. Finally, for (K$\diamond$7), let K be complete and $\epsilon \in K \diamond (\alpha \vee \beta)$. Then by Levi identity and (K$-\!\!\diamond\!\!-$4), $\epsilon \in (K -\!\!\diamond\!\!- (\neg\alpha \wedge \neg\beta)) + (\alpha \vee \beta)$. By deduction $(\alpha \vee \beta) \rightarrow \epsilon \in K -\!\!\diamond\!\!- (\neg\alpha \wedge \neg\beta)$, from which it follows by (K$-\!\!\diamond\!\!-$7) that $(\alpha \vee \beta) \rightarrow \epsilon \in Cn((K -\!\!\diamond\!\!- \neg\alpha) \cup (K -\!\!\diamond\!\!- \neg\beta))$. Then $\epsilon \in Cn((K -\!\!\diamond\!\!- \neg\alpha) + \alpha \cup (K -\!\!\diamond\!\!- \neg\beta + \beta))$. Hence by Levi identity, $\epsilon \in Cn(K \diamond \alpha \cup K \diamond \beta)$. ■

[2] A partial proof of this theorem was provided in [7].

Proof of Theorem 2 ($K{-\!\!\!\diamond\!\!\!-}0$), ($K{-\!\!\!\diamond\!\!\!-}1$), ($K{-\!\!\!\diamond\!\!\!-}3$), ($K{-\!\!\!\diamond\!\!\!-}4$), ($K{-\!\!\!\diamond\!\!\!-}R$), and ($K{-\!\!\!\diamond\!\!\!-}8$) follow trivially. For ($K{-\!\!\!\diamond\!\!\!-}5$), let $\epsilon \in K{-\!\!\!\diamond\!\!\!-}(\alpha) \cap (K{-\!\!\!\diamond\!\!\!-}\beta)$. We need to prove that $\epsilon \in K{-\!\!\!\diamond\!\!\!-}(\alpha \wedge \beta)$. It follows by Harper identity that $\epsilon \in (K \diamond \neg\alpha) \cap K \cap (K \diamond \neg\beta) \cap K$. Due to $\neg\alpha \equiv (\neg\alpha \vee \neg\beta) \wedge \neg\alpha$ and $\neg\beta \equiv (\neg\alpha \vee \neg\beta) \wedge \neg\beta$, we obtain by (K$\diamond$4) that $\epsilon \in K \diamond ((\neg\alpha \vee \neg\beta) \wedge \neg\alpha)$ and $\epsilon \in K \diamond ((\neg\alpha \vee \neg\beta) \wedge \neg\beta)$. (K$\diamond$5) yields $\epsilon \in K \diamond (\neg\alpha \vee \neg\beta) + \neg\alpha$ and $\epsilon \in K \diamond (\neg\alpha \vee \neg\beta) + \neg\beta$. By deduction and (K$\diamond$0), $\neg\alpha \rightarrow \epsilon \in K \diamond (\neg\alpha \vee \neg\beta)$ and $\neg\beta \rightarrow \epsilon \in K \diamond (\neg\alpha \vee \neg\beta)$, from which it follows by (K$\diamond$0) that $(\neg\alpha \vee \neg\beta) \rightarrow \epsilon \in K \diamond (\neg\alpha \vee \neg\beta)$, and due to (K$\diamond$0) and (K$\diamond$1) $\epsilon \in K \diamond (\neg\alpha \vee \neg\beta)$. Then $\epsilon \in K \diamond (\neg\alpha \vee \neg\beta) \cap K$, hence by (K$\diamond$4) and Harper identity $\epsilon \in K{-\!\!\!\diamond\!\!\!-}(\alpha \wedge \beta)$. For ($K{-\!\!\!\diamond\!\!\!-}6$), let $\beta \rightarrow \alpha \in K{-\!\!\!\diamond\!\!\!-}\alpha$ and $\alpha \rightarrow \beta \in K{-\!\!\!\diamond\!\!\!-}\beta$. It follows by Harper that $\beta \rightarrow \alpha \in K \diamond \neg\alpha \cap K$ and $\alpha \rightarrow \beta \in K \diamond \neg\beta \cap K$. By (K$\diamond$0) and (K$\diamond$1) $\neg\beta \in K \diamond \neg\alpha$ and $\neg\alpha \in K \diamond \neg\beta$. By (K$\diamond$6), $K \diamond \neg\alpha = K \diamond \neg\beta$. Then $K \diamond \neg\alpha \cap K = K \diamond \neg\beta \cap K$. Hence, by Harper identity, $K{-\!\!\!\diamond\!\!\!-}\alpha = K{-\!\!\!\diamond\!\!\!-}\beta$. For ($K{-\!\!\!\diamond\!\!\!-}7$), let K be complete and let $\epsilon \in K{-\!\!\!\diamond\!\!\!-}(\alpha \wedge \beta)$. We need to prove that $\epsilon \in Cn((K{-\!\!\!\diamond\!\!\!-}\alpha) \cup (K{-\!\!\!\diamond\!\!\!-}\beta))$. By Harper identity, $\epsilon \in K \diamond (\neg\alpha \vee \neg\beta) \cap K$. From $\epsilon \in K \diamond (\neg\alpha \vee \neg\beta)$, it follows by (K$\diamond$7) that $\epsilon \in Cn(K \diamond \neg\alpha \cup K \diamond \neg\beta)$. By Harper identity, $\epsilon \in Cn((K{-\!\!\!\diamond\!\!\!-}\alpha \cap K) \cup (K{-\!\!\!\diamond\!\!\!-}\beta \cap K)) \subseteq Cn(K{-\!\!\!\diamond\!\!\!-}\alpha \cup K{-\!\!\!\diamond\!\!\!-}\beta)$. ∎

Proof of Theorem 3 We give a proof sketch here. To one direction, let ${-\!\!\!\diamond\!\!\!-}$ defined as $[[K{-\!\!\!\diamond\!\!\!-}\alpha]] = (\bigcup_{\omega_i \in [[K]]} \min([[\neg\alpha]], \leq_{\omega_i})) \cup [[K]]$. Then we need to prove that ${-\!\!\!\diamond\!\!\!-}$ satisfies the erasure postulates.

To the other direction, let ${-\!\!\!\diamond\!\!\!-}$ be a erasure operator as defined in Definition 2 and let us consider the assignment associating every world ω_i to a relation $\leq_{\omega_i}$ such that for all worlds $\omega, \omega' \in \Omega$ it follows that $\omega \leq_{\omega_i} \omega'$ if and only if $\omega = \omega_i$ or $[[K_{\omega_i}{-\!\!\!\diamond\!\!\!-}\neg\alpha_{\omega,\omega'}]] = \{\omega, \omega_i\}$. Then we need to show that $\leq_{\omega_i}$ is a partial preorder (reflexive, transitive). Then we prove that the assignment is faithful, i.e., given $\omega \neq \omega_i$, we need to prove that $\omega_i <_{\omega_i} \omega$. Finally we need to prove that $[[K{-\!\!\!\diamond\!\!\!-}\alpha]] = (\bigcup_{\omega_i \in [[K]]} \min([[\neg\alpha]], \leq_{\omega_i})) \cup [[K]]$. ∎

References

1. Alchourrón, C., Gärdenfors, P., Makinson, D.: On the logic of theory change: partial meet contraction and revision functions. J. Symb. Log. **50**, 510–530 (1985)
2. Eiter, T., Gottlob, G.: On the complexity of propositional knowledge base revision, updates, and counterfactuals. Artif. Intell. **57**, 227–270 (1992)
3. Fermé, E., Gonçalves, S.: On the logic of theory change iteration of km-update. Int. J. Approx. Reas. **162**, 109005 (2023)
4. Fermé, E., Konieczny, S., Pérez, R.P., Schwind, N.: Credible models of belief update. In: 20th International Conference on Principles of Knowledge Representation and Reasoning (KR-2023), pp. 252–261. International Joint Conferences on Artificial Intelligence Organization (2023)
5. Harper, W.L.: Ramsey test conditionals and iterated belief change. In: Harper, W.L., Hooker, C.A. (eds.) Foundations of Probability Theory, Statistical Inference, and Statistical Theories of Science, vol. 1, pp. 117–136. D. Reidel Publishing Co., Dordrecht (1976)
6. Herzig, A., Rifi, O.: Propositional belief base update and minimal change. Artif. Intell. **115**(1), 107–138 (1999)

7. Katsuno, H., Mendelzon, A.: On the difference between updating a knowledge base and revising it. In: Gärdenfors, P. (ed.) Belief Revision. Cambridge Tracts in Theoretical Computer Science, vol. 29, pp. 183–203. Cambridge University Press (1992)
8. Keller, A.M., Winslett, M.: On the use of an extended relational model to handle changing incomplete information. IEEE Trans. Softw. Eng. **11**(7), 620–633 (1985)
9. Lang, J.: Belief update revisited. IJCAI **7**, 6–12 (2007)
10. Levi, I.: Subjunctives, dispositions, and chances. Synthese **34**, 423–455 (1977)
11. Peppas, P., Williams, M.A.: Constructive modelings for theory change. Notre Dame J. Formal Logic **36**(1), 120–133 (1995). https://doi.org/10.1305/ndjfl/1040308831
12. Slota, M., Leite, J.: On semantic update operators for answer-set programs. In: ECAI 2010, pp. 957–962. IOS Press (2010)
13. Winslett, M.: Reasoning about action using a possible models approach. In: AAAI, pp. 89–93 (1988)

Stable Reasoning, ASP and the Interrogative Model of Inquiry

David Pearce and Agustín Valverde

Abstract In this note we explore the relation between question-answering in answer set programming and the interrogative model of scientific inquiry proposed by Jaakko Hintikka. To this end we extend the fixpoint property of propositional equilibrium models to the first order case. In this manner we obtain a rather close agreement between the interrogative model and question-answering in ASP; this may be relevant to the topic of explanatory ASP.

1 Introduction

What we may call *stable reasoning* comes in several varieties that share some common features. One is a very general property of stable reasoning that manifests itself in different ways. Roughly, it is the idea that a stable reasoner extends their initial assumptions, beliefs or theory in such a way that it becomes complete and also in a sense 'protected'. This is clearly visible in Dung's argumentation theory, where a stable extension of an argumentation system is one that is complete, conflict-free and moreover attacks every argument that does not belong to it [1]. A similar idea is also present in Stalnaker's concept of *stable belief set* [8] with the requirement that for any sentence φ not in a belief set S, $\neg B\varphi \in S$, where B is the belief operator. This later became a design feature when defining nonmonotonic modal logics of knowledge and belief.

D. Pearce
Universidad Politécnica de Madrid, Madrid, Spain
e-mail: david.pearce@upm.es

A. Valverde (✉)
Universidad de Málaga, Malaga, Spain
e-mail: a_valverde@uma.es

L. T. Kóczy et al. (eds.), *Computational Intelligence and Mathematics for Tackling Complex Problems 6*, Studies in Computational Intelligence 1222,
https://doi.org/10.1007/978-3-031-97879-1_16

For stable models of logic programs, we have the idea of *completion* expressed in Proposition 1 below. Let HT be the propositional logic of *here-and-there*.[1] Its truth relation is denoted by $\models$ and its consequence operator by Cn_{HT}. As is well-known, HT has a nonmonotonic extension, called *equilibrium logic*, whose models correspond to the stable models or answer sets of logic programs [7]. In this way HT and equilibrium logic provide a foundation for reasoning about answer set programs.

Definition 1 *Let Π be a propositional theory. A set E of formulas extending Π is said to be a* completion *of Π iff $E = Cn_{HT}(\Pi \cup \{\neg\varphi : \varphi \notin E\})$.*

Equilibrium models (and therefore stable models) correspond precisely to completions. For any model $\mathcal{M}$, set $Th(\mathcal{M}) = \{\varphi : \mathcal{M} \models \varphi\}$.

Proposition 1 [6] *If Π is a propositional theory, E is a completion of Π iff $E = Th(\mathcal{M})$ for some equilibrium model $\mathcal{M}$ of Π. Similarly, $\mathcal{M} = \langle T, T\rangle$ is an equilibrium model of Π iff $T = E \cap At$, for some completion E of Π. (Where At denotes the set of all atoms in the language of Π.)*

For logic programs, the aspect of 'protected' arises from the basic definition of stability in the form of equilibrium models (see appendix). In this framework, any total model of the theory is already complete. But besides this, the equilibrium construction ensures that for that particular total model the theory does not admit any nontotal 'submodel'. There is no other model of the theory that undermines the stable one by having more uncertainty.

These different forms of stability are not merely analogous, but are closely connected mathematically. Already Dung [1] explored the links between argumentation theory and stable models of logic programs, while many scholars have established connections between stable models and nonmonotonic modal logics (see e.g. [5]).

There is another feature of stable reasoning that is more methodological in spirit and which emerges most clearly in answer set programming (ASP). It is the idea that an ASP program describes a particular domain, possibly in a very general way, while a stable model of the program can solve a particular problem by representing the 'actual' world or state of affairs for some specific instance given by a set of initial conditions. Even though an ASP solver is not usually considered to be 'goal-directed', nevertheless we can consider the ASP methodology to be problem-oriented. In terms of logical methodology we can describe it as a problem-solving approach, in contrast to, let us say, a theorem-proving one. In the latter case, we take a theory and make inferences from it. In the case of ASP we describe a domain and search for a particular model that represents a state of affairs or perhaps a plan of action, and thereby resolves a problem that concerns us. So, ultimately, ASP can and does answer questions. Before we explore this feature of question-answering more closely, let us recast Proposition 1 in a slightly different form.

[1] Notation, terminology and technical details on HT, its first-order extension, QHT, and equilibrium logic are collected in the appendix. We assume the reader has some familiarity with answer set programming and the stable model semantics.

If $\mathcal{M} = \langle H, T \rangle$ is an HT interpretation, we denote by $\mathcal{D}(\mathcal{M})$ the diagram of $\mathcal{M}$, i.e. the collection of literals true in $\mathcal{M}$. So in particular an atom $a \in \mathcal{D}(\mathcal{M})$ if $a \in H$, and a literal $\neg a \in \mathcal{D}(\mathcal{M})$ if $a \notin T$. As usual $Th(\mathcal{M})$ denotes the collection of all formulas true in $\mathcal{M}$.

Proposition 2 *Let Π be a propositional theory. $\mathcal{M}$ is an equilibrium model of Π if and only if*

$$Th(\mathcal{M}) = Cn_{HT}(\Pi \cup \{\neg l : l \notin \mathcal{D}(\mathcal{M})\}). \tag{1}$$

It is important to bear in mind that equation (1) refers to inference in the nonclassical logic, HT. In this logic the law of excluded middle does not hold. This means that if a is an atom not in $\mathcal{D}(\mathcal{M})$ then $\neg a$ belongs to the theory of $\mathcal{M}$. But if l is a negative literal, say $\neg b$, then (1) ensures that $\neg\neg b \in Th(\mathcal{M})$.

2 Fixpoints for Quantified Equilibrium Logic

What happens if we look for an analogous property of equilibrium models in the first order case? In this case equilibrium models are defined using the quantified version of HT logic; we denote this by QHT and its consequence relation by Cn_{QHT}. This provides a (stable model) semantics for first order theories and for logic programs with variables [7].

For quantified equilibrium logic we are going to split the properties expressed in Propositions 1 and 2 into two parts. In one case we deal with completions and no restrictions on the semantics, in the other case we will use a particular version of the QHT semantics.

Let us reconsider Proposition 1. In one direction the analogous property of completion holds for first order equilibrium models as follows.

Proposition 3 *Let Π be a first order theory in L and let $\mathcal{M} = \langle (D, \sigma), H, T \rangle$ be a QHT model. Suppose that $Th(\mathcal{M}) = Cn_{QHT}(\Pi \cup \{\neg\varphi \in L(D) : \varphi \notin Th(\mathcal{M})\})$. Then $\mathcal{M}$ is an equilibrium model of Π.*

For the converse property we would like to sharpen the conditions. To do this, however, we need to restrict the semantics. In the general case where we have arbitrary theories we restrict to the standard name assumption, SNA (see Appendix). Although restrictive, SNA it is not an unusual assumption when dealing with logic programs with variables. A common practice is to restrict the semantics to the Herbrand universe and Herbrand models, thereby obeying SNA. However, in the case of universal theories or logic programs we can actually establish a similar fixpoint condition under the less restrictive PNA, the parameter name assumption.

Let Π be a first order theory or logic program in L and let $\mathcal{M} = \langle (D, \sigma), H, T \rangle$ be a QHT model for L. We denote by $\mathcal{D}(\mathcal{M})$ the diagram of $\mathcal{M}$; in this case it is the collection of all literals in $L(D)$ that are true in $\mathcal{M}$. Here are our partial counterparts to Proposition 2.

Proposition 4 *(i) Assume SNA semantics for QHT and let* Π *be a first order theory. Let* $\mathcal{M}$ *be an equilibrium model of* Π*. Then*

$$Th(\mathcal{M}) = Cn_{QHT}(\Pi \cup \{\neg l \in L(D) : l \notin \mathcal{D}(\mathcal{M})\}). \tag{2}$$

(ii) Assume PNA semantics for QHT and let Π *be a universal theory*[2] *or logic program. Then Eq. (2) holds.*

Proof. For reasons of space we show only part (i). The proof of part (ii) is different and relies on the fact that in QHT universal formulas are preserved under submodels.

Let $\mathcal{M} = \langle (D, \sigma), T, T\rangle$ be an equilibrium model of Π; if we assume the SNA semantics for QHT, $\sigma : C \to D$ is a bijection. We show that

$$Th(\mathcal{M}) = Cn_{QHT}(\Pi \cup \{\neg l \in L(D) : l \notin \mathcal{D}(\mathcal{M})\})$$

In the sequel, we are going to use the following notation. If l is an atom with parameters in D, then $\sigma^{-1}(l)$ is the atom obtained by replacing every parameter d by the constant $\sigma^{-1}(d)$. Analogously, if l is an atom in L, then $\sigma(l)$ is the atom obtained by replacing every every constant c by the parameter $\sigma(c) \in D$.

Let l be a literal in L such that $\mathcal{M} \not\models l$, then $\mathcal{M} \models \neg l$. That is, $\mathcal{M} \models \Pi \cup \{\neg l \in L(D) : l \notin \mathcal{D}(\mathcal{M})\}$ and therefore $Cn_{QHT}(\Pi \cup \{\neg l \in L(D) : l \notin \mathcal{D}(\mathcal{M})\} \subseteq Th(\mathcal{M}))$.

For the other inclusion, suppose that

$$\mathcal{M}' = \langle (D', \sigma'), H', T'\rangle \models \Pi \cup \{\neg l \in L(D) : l \notin \mathcal{D}(\mathcal{M}))\}$$

Trivially, this is equivalent to

$$\mathcal{M}'' = \langle (D, \sigma), \sigma \circ \sigma'^{-1}(H'), \sigma \circ \sigma'^{-1}(T')\rangle \models \Pi \cup \{\neg l \in L(D) : l \notin \mathcal{D}(\mathcal{M}))\}$$

since $\sigma : C \to D$ and $\sigma' : C \to D'$ are both bijections.

If $\sigma(a) \in T$ then $\neg a \notin \mathcal{D}(\mathcal{M})$ and $\mathcal{M}'' \models \neg\neg a$; that is, $\mathcal{M}'', t \models a$ and $\sigma'(a) \in T'$. On the other hand, if $\sigma'(a) \in T'$ then $\mathcal{M}' \not\models \neg a$ and necessarily $\mathcal{M} \models a$ and $\sigma(a) \in T$. Therefore, $T = \sigma \circ \sigma'^{-1}(T')$ and, since $\mathcal{M}$ is an equilibrium model, $\sigma \circ \sigma'^{-1}(H') = T$ and $\mathcal{M} = \mathcal{M}''$. Consequently, $\mathcal{M}'' \models Th(\mathcal{M})$ and equivalently $\mathcal{M}' \models Th(\mathcal{M})$.$\dashv$

3 Equilibrium Logic and the Interrogative Model

Beginning in the early 1960s, philosophers of science started to doubt the adequacy of the prevalent logical empiricist programme of rational reconstruction of scientific knowledge based on the concepts and methods of classical logic in the Frege-Hilbert

---

[2] That is a theory all of whose formulas are in prenex form and contain only universal quantifiers.

tradition. This led a number of methodologists to question what was commonly known as the 'received view' of scientific theories and their methods of explanation. A prominent critic of the standard view was the American philosopher, Larry Laudan, who stressed that first and foremost scientific inquiry is a problem-solving and question-answering process [4]. Laudan discussed many aspects of scientific progress, but did not provide a formal, logical model to fit the activity of problem-solving.

One such attempt was made by Jaakko Hintikka, who sketched the idea of an interrogative model of inquiry and proposed a formal analysis of it in classical logic. This was intended to reflect the question-answering process.[3] Hintikka's proposal is that we should start from a general theory, Π, reconstructed in an appropriate logic, with the idea that the actual world is accurately described by some particular model, say $\mathcal{M}$, of the theory. When an inquirer wishes to solve a problem expressed by a sentence say φ in the language of Π (suitably extended by constant names), they look not for the truth of φ in Π, but rather whether Π together with some set of atomic or negated atomic sentences true in $\mathcal{M}$ can derive φ. Formally, the characterisation is

$$\Pi \vdash \{l_1, l_2, \ldots, l_n\} \rightarrow \varphi \tag{3}$$

where $l_1, l_2, \ldots, l_n$ are literals true in $\mathcal{M}$. If the logic of $\vdash$ is compact, then (3) can be equivalently rewritten using the diagram $\mathcal{D}(\mathcal{M})$ of $\mathcal{M}$:

$$\Pi \cup \mathcal{D}(\mathcal{M}) \vdash \varphi \tag{4}$$

When these formulas hold, Hintikka calls φ a *model consequence* of Π in $\mathcal{M}$. Clearly if $\mathcal{M}$ is completely known, then we could check the truth value of φ in $\mathcal{M}$ directly. So the thought behind Hintikka's proposal seems to be that with the help of the theory Π we may acquire knowledge about some general property of $\mathcal{M}$ using the theory plus a set of literals true in $\mathcal{M}$.

Following Hintikka we may consider a subclass of models of a theory Π to be *optimal* if for any model $\mathcal{M}$ in that class and any problem φ in the language of Π (plus constants for $\mathcal{M}$) either φ or $\neg\varphi$ is a model consequence of Π in $\mathcal{M}$. In other words, question-answering is complete for any model in the class. As a limiting case the theory itself may be considered optimal if the class of all its models is optimal. As Hintikka observed, in classical first order logic this property of optimality coincides with model-completeness.[4]

It is instructive to explore some of the parallels between Hintikka's interrogative model and question-answering in the style of ASP. First, let us consider propositional logic and note that if Π has the form of a (ground) disjunctive logic program, then

---

[3] Hintikka wrote several articles on the interrogative model, but we will focus on [3]. Here the logic is not specified but some properties and applications of the model are described for standard first order logic.

[4] It is well known that in classical, first order logic, completeness and model-completeness are different, non-equivalent properties. So this is a distinguishing feature of the interrogative model.

the property (1) holds where $\mathcal{D}(\mathcal{M})$ is just the *positive* diagram of $\mathcal{M}$, in other words the set of atoms true in $\mathcal{M}$. Both HT and QHT are compact and so for a disjunctive program Π, any formula true in a stable model $\mathcal{M}$ of Π is a model consequence of Π in $\mathcal{M}$ is the sense of Hintikka. In fact, for any φ true in $\mathcal{M}$, there is a finite set of *negated atoms* $l_1, l_2, \ldots, l_n$ such that

$$\Pi \vdash_{HT} \{l_1, l_2, \ldots, l_n\} \rightarrow \varphi. \tag{5}$$

When Π is a general propositional theory, by Proposition 2, (5) holds for a set $\{l_1, l_2, \ldots, l_n\}$ of negated literals. So in Hintikka's terms, for any theory the collection of its stable models is in a sense optimal with respect to question answering.

Since the interrogative model is formulated in first order logic we obtain a closer relation to ASP if we move to quantified equilibrium logic. Then if Π is a logic program with variables in language L, and we assume PNA, for any L-sentence φ, if φ is true in an equilibrium model $\mathcal{M}$ of Π, by Proposition 4 and the property of compactness there is a finite set of *negated literals* $l_1, l_2, \ldots, l_n$ such that

$$\Pi \vdash_{QHT} \{l_1, l_2, \ldots, l_n\} \rightarrow \varphi. \tag{6}$$

This shows that for programs with variables, stable or equilibrium models are also optimal with respect to question-answering. Since equilibrium models are complete, for any L-sentence φ, either φ or $\neg\varphi$ is derivable from Π together with a finite set of (negative) assumptions, and so amounts to a model consequence of Π, adapting Hintikka's terminology.

These two interrogative styles are remarkably similar. The fact that in (6) for example we have negated literals rather than simply literals is a reflection of the fact that a nonclassical logic underlies the programming approach of ASP. In classical logic a set of negated literals would just be a set of literals *simpliciter*.

The other difference arises from the idea in ASP that the assumptions that one adds to a program to derive a certain consequence are in negative form. In the setting of logic programming, negation is often thought to mirror the idea of a default. Viewed in this way, even in the first order setting, any sentence true in an answer set can be derived from the program (in the nonclassical base logic) merely by adding a finite set of default assumptions.

4 Conclusions

We have extended some known fixpoint properties of equilibrium and stable models from the propositional to the first order case. This allows us to compare question-answering in ASP with the interrogative model of inquiry proposed by Jaakko Hintikka over 40 years ago. We noted a remarkable similarity between the two forms of reasoning; the main differences arising from the special role that negation and default assumptions play in ASP and from the fact that its logical basis is nonclassical. Allowing for these differences, we can say that stable models are optimal with respect to question-answering.

This comparison may be relevant to the topic of AI and accountability that has recently become a very lively area of discussion and investigation. Within this area explanatory ASP is an active topic of research whose aim is to enhance logic-based systems such as ASP with tools that can provide users with explanations for the solutions and outcomes of such systems in practical applications (see e.g. [2]).

Hintikka's approach to describing the process of inquiry and understanding is clearly different from the deductive-nomological model of explanation often invoked in the study of the natural and social sciences. By comparison, in the sciences of the artificial, relevant concepts of explanation and understanding have received less attention until now. Hintikka's model of inquiry, suitably adapted to theories and contexts relevant to AI, may provide useful insights for explanatory AI in the case of logic-based systems. Here is one example where our conclusions may be relevant to an issue in the area of explanatory ASP: they may help to understand why in some approaches to explanatory ASP, such as the system xClingo [2], one does not seek to explain negated atoms.

Acknowledgements Supported by project LIANDA—BBVA Foundation Grants for Scientific Research Projects, and by Junta de Andalucía (Spain) project TIC115.

Appendix: Equilibrium Logic and the Logic of Here-and-There

For a description of equilibrium logic, the quantified logic of here-and-there, QHT, and their relation to answer set programming and stable model semantics, see [7]. Here is a brief summary. Let $L = (C, P)$ be a first order language with set of constants C and predicates P. The QHT semantics is a form of possible worlds semantics for first order intuitionistic logic [9] where all domains are equal and there are just two nodes or 'worlds', h and t, with $h \leq t$. We can represent a QHT interpretation for L as a triple $\langle (D, \sigma), H, T \rangle$, where D is the domain of objects, σ is an interpretation of the language L in the domain D. We denote by $L(D)$ the extension of L by constants for each element of D and we will identify σ with its extension to $L(D)$. Formally we have

- $\sigma : C \cup D \to D$ verifies $\sigma(d) = d$ if $d \in D$, and
- H, T are sets of atomic formulas in $L(D)$, such that $H \subseteq T$.

The *satisfaction* relation for a QHT-interpretation $\mathcal{M} = \langle (D, \sigma), H, T \rangle$, at a node $w \in \{h, t\}$, with $h \leq t$, is defined as follows:

- $\mathcal{M}, h \models p(t_1, \dots)$ if $p(\sigma(t_1), \dots) \in H$,
- $\mathcal{M}, t \models p(t_1, \dots)$ if $p(\sigma(t_1), \dots) \in T$,
- $\mathcal{M}, w \models \varphi \wedge \psi$ if $\mathcal{M}, w \models \varphi$ and $\mathcal{M}, w \models \psi$,
- $\mathcal{M}, w \models \varphi \vee \psi$ if $\mathcal{M}, w \models \varphi$ or $\mathcal{M}, w \models \psi$,

- $\mathcal{M}, w \models \varphi \to \psi$ if, for each node w' such that $w \le w'$, $\mathcal{M}, w' \not\models \varphi$ or $\mathcal{M}, w' \models \psi$,
- $\mathcal{M}, w \models \neg\varphi$ if $\mathcal{M}, t \not\models \varphi$
- $\mathcal{M}, w \models \forall x\varphi(x)$ if $\mathcal{M}, w \models \varphi(d)$ for every $d \in D$.
- $\mathcal{M}, w \models \exists x\varphi(x)$ if $\mathcal{M}, w \models \varphi(d)$ for some $d \in D$.

As usual a *literal* is an atomic formula or its negation. A sentence φ is *true* in an L interpretation $\mathcal{M}$, in symbols $\mathcal{M} \models \varphi$, if $\mathcal{M}, h \models \varphi$. For any set Γ of sentences and any sentence φ, we write $\Gamma \models \varphi$ if every QHT-interpretation satisfying all formulas in Γ satisfies φ also.

Let $\mathcal{M} = \langle (D, \sigma), H, T \rangle$ be a model of a theory Π. We say that: $\mathcal{M}$ is a PNA-model (parameter names assumption model) if the restriction of σ to C is surjective; $\mathcal{M}$ is a UNA-model (unique names assumption model) if the restriction of σ to C is injective; and we say that $\mathcal{M}$ is a SNA-model (standard names assumption model) if the restriction of σ to C is a bijection.

If $H = T$, we say that the model in question is *total*. If $\mathcal{M} = \langle (D, \sigma), T, T \rangle \models \Pi$ and if as well $\langle (D, \sigma), H, T \rangle \not\models \Pi$ for every $H \subset T$, we say that $\mathcal{M}$ is an *equilibrium* model of Π.

For the *propositional logic of here-and-there*, HT, we take a propositional language L and regard an HT interpretation for L as a pair $\mathcal{M} = \langle H, T \rangle$ of sets of L-atoms, with $H \subseteq T$. Again we can think of this structure as comprising the sets of atoms true at two nodes h and t. And, if, as before, $w \in \{h, t\}$, the above satisfaction clauses for $\wedge, \vee, \to, \neg$ define the truth relation for any L-formula φ in $\mathcal{M}$. *Equilibrium models* for HT are defined in the same way as before, i.e. they are total models $\langle T, T \rangle$ of a theory Π such that there is no HT model $\langle H, T \rangle$ of Π such that $H \subset T$. The *logics* HT and QHT are determined by their respective *valid* formulas, i.e. those true in every HT or QHT interpretation. The inference relations of HT and QHT are denoted by $\vdash_{HT}$ and $\vdash_{QHT}$ respectively.

The main connection between equilibrium logic and ASP is the formal equivalence between answer sets and equilibrium models [7].

References

1. Dung, P.M.: On the acceptability of arguments and its fundamental role in nonmonotonic reasoning and logic programming. In: Bajcsy, R. (ed.) Proceedings of the 13th International Joint Conference on Artificial Intelligence, pp. 852–859. Morgan Kaufmann (1993)
2. Fandinno, J., Cabalar, P., Muñiz, B.: A system for explainable answer set programming. Electron. Proc. Theor. Comput. Sci. **325**, 124–136 (2020)
3. Hintikka, J.: The logic of science as a model-oriented logic. In: Asquith, P.D., Kitcher, P. (eds.) PSA 1984, vol. 1, pp. 177 – 185 (1984)
4. Laudan, L.: Progress and Its Problems. University of California Press (1977)
5. Marek, W., Truszczynski, M.: Nonmonotonic Reasoning: Context-Dependent Reasoning. Springer (1993)
6. Pearce, D.: From here to there: stable negation in logic programming. In: Gabbay, D., Wansing, H. (eds.) What Is Negation? Kluwer (1999)

7. Pearce, D., Valverde, A.: Quantified equilibrium logic and foundations for answer set programs. In: Garcia de la Banda, M., Pontelli, E. (eds.) Proceedings ICLP 2008. LNCS, vol. 5366, pp. 546–560. Springer (2008)
8. Stalnaker, R.: A note on non-monotonic modal logic. Artif. Intell. **64**(2), 183–196 (1993)
9. van Dalen, D.: Logic and Structure. Universitext. Springer, London (2012)

Hypergraph Logic Program Representation Versus Stratified Programs

David Lobo, Jesús Medina, José R. Portillo, and José A. Torné-Zambrano

Abstract Multi-adjoint normal logic programming is a general non-monotonic logic programming framework, which makes it ideal for modeling complex scenarios. Hypergraph representation has been proved to be an appropriate tool in the study of different properties of a logic program, whereas the use of a stratification has provided interesting results related to the existence and unicity of stable models. In this paper, we will see the relation between the p-condensation graph of a program and its "optimal" stratification.

Keywords Logic programming · Hypergraphs · Negation operator

1 Introduction

Logic programming arises as a mathematical tool for expressing and solving problems through declarative specifications, becoming particularly powerful when dealing with knowledge-based systems and automated reasoning [14, 18]. Among all the generalizations of logic programming, it stands out the multi-adjoint normal one (MANLP) [4, 5], allowing to use different adjoint pair in the construction of the

D. Lobo · J. Medina · J. A. Torné Zambrano (✉)
Department of Mathematics, University of Cádiz, Cádiz, Spain
e-mail: joseantonio.torne@uca.es

D. Lobo
e-mail: david.lobo@uca.es

J. Medina
e-mail: jesus.medina@uca.es
URL: https://mcis.uca.es/

J. R. Portillo
Department of Applied Mathematics 1, University of Sevilla, Seville, Spain
e-mail: josera@us.es
URL: https://ma1.us.es; https://imus.us.es

University Institute for Research in Mathematics (IMUS), Seville, Spain

L. T. Kóczy et al. (eds.), *Computational Intelligence and Mathematics for Tackling Complex Problems 6*, Studies in Computational Intelligence 1222,
https://doi.org/10.1007/978-3-031-97879-1_17

rules of the programs [6], together with less restrictive properties in the considered connectives and the use of negation operators, providing a more flexible framework.

Taking into account that one of the main demanding tasks in MANLP is the study of conditions to ensure the existence and the unicity of stable models [11, 15], we can understand the importance of stratifications in logic programming. Indeed, the determination of stratified logic programs is a relevant sufficient condition in order to determine whether a logic program has a (unique) stable model [1, 17].

Furthermore, multi-adjoint logic programs have been represented by hypergraphs obtaining interesting results. For example, this relationship was used in [8, 9] to the termination property of the immediate consequences operator. This representation was extended in [10].

In this paper, we will provide a hypergraph representation of a MANLP completely characterizing the considered program, complementing the one given in [9, 10]. Moreover, we will show that from the p-condensation graph of the program $\mathbb{P}$, we can define an "optimal" stratification of the given program. As a consequence, it also makes possible to compute the least model of the subprogram associated with each stratum in parallel.

2 Multi-adjoint Normal Logic Programming

The algebraic structure considered in this framework is a local multi-adjoint normal Σ-algebra, which considers a bounded lattice endowed with a negation operator and adjoint pairs, which generalize left-continuous t-norms and their residuated implications. For more details, the reader can see [4, 16].

In this paper, we will consider a Σ-algebra $\mathfrak{L}$ based on a multi-adjoint lattice [7] $(L, \preceq, \leftarrow_1, \&_1, \dots, \leftarrow_n, \&_n)$, a negation operator $\neg$, and other operators, such as conjunctors $\wedge_1, \dots, \wedge_k$, disjunctors $\vee_1, \dots, \vee_l$ and general aggregators $@_1, \dots, @_h$. Considering the multi-adjoint Σ-algebra $\mathfrak{L}$, a countable set of propositional symbols Π and the algebra of well-formed formulas $\mathfrak{F}$, we introduce the definition of multi-adjoint normal logic program [4, 16].

Definition 1 A *multi-adjoint normal logic program*, MANLP in short, is a set of rules of the form $\langle A \leftarrow_i \mathcal{B}, \vartheta\rangle$ such that:

1. The *rule* $A \leftarrow_i \mathcal{B}$ is a formula of $\mathfrak{F}$.
2. The *confidence factor* ϑ is an element (a truth-value) of L.
3. The *head* of the rule A is a propositional symbol of Π.
4. The *body* formula $\mathcal{B}$ is a formula of $\mathfrak{F}$ of the form $@[B_1, \dots, B_s, \neg B_{s+1}, \dots, \neg B_r]$ built from propositional symbols $B_1, \dots, B_r$ ($r \geq 0$, $B_i \neq B_j$, for $i \neq j$) by the use of conjunctors $\&_1, \dots, \&_n$ and $\wedge_1, \dots, \wedge_k$, disjunctors $\vee_1, \dots, \vee_l$, aggregators $@_1, \dots, @_m$ and elements of L (which composition is represented by $@$).
5. *Facts* are rules with the body $\top$.

Example 1 In this example in the unit interval $L = [0, 1]$, the adjoint pairs corresponding to the product, Gödel and Lukasiewicz t-norms, $(\&_P, \leftarrow_P)$, $(\&_G, \leftarrow_G)$, $(\&_Ł, \leftarrow_Ł)$ are considered, together with the weighted sums $@_{(2,1)}$ and $@_{(1,3)}$ defined as $@_{(2,1)}(x, y) = \dfrac{2x + y}{3}$ and $@_{(1,3)}(x, y) = \dfrac{x + 3y}{4}$, for every $(x, y) \in [0, 1]^2$. Moreover, the negation $\neg$, defined as $\neg(x) = 1 - x$ for $x \in [0, 1]$, will also be taken into account in the program. Specifically, the following normal program $\mathbb{P}$ will be analyzed in the rest of the paper:

$$
\begin{array}{lll}
\langle c \leftarrow_P n \,\&_P \neg u, 0.8\rangle & \langle p \leftarrow_P n, 0.9\rangle & \langle t \leftarrow_P 1.0, 0.1\rangle \\
\langle n \leftarrow_P c, 0.8\rangle & \langle m \leftarrow_P n \,\&_G a, 0.8\rangle & \langle n \leftarrow_P 1.0, 0.5\rangle \\
\langle n \leftarrow_P @_{(2,1)}(\neg f, \neg h), 0.6\rangle & \langle f \leftarrow_P u, 0.9\rangle & \langle u \leftarrow_P 1.0, 0.2\rangle \\
\langle h \leftarrow_P f, 0.7\rangle & \langle a \leftarrow_P @_{(1,3)}(\neg f, \neg u), 1.0\rangle & \langle u \leftarrow_P 1.0, 0.2\rangle \\
\langle u \leftarrow_G h \,\&_Ł f, 0.7\rangle & &
\end{array}
$$

Note that, in non-commutative aggregators such as $@_{(1,3)}$ or $@_{(2,1)}$, the order of variables in the input follows the enumeration defined by the set Π. In our example, we must write $@_{(1,3)}(\neg f, \neg u)$ instead of $@_{(3,1)}(\neg u, \neg f)$.

The semantics is given by the notion of model:

Definition 2 Given a mapping (called interpretation) $I \colon \Pi \to L$, a weighted rule $\langle A \leftarrow_i \mathcal{B}, \vartheta\rangle$ is *satisfied* by I, if $\vartheta \preceq \hat{I}(A \leftarrow_i \mathcal{B})$, where $\hat{I}$ is the unique homomorphic extension of I on $\mathfrak{F}$. An interpretation $I \colon \Pi \to L$ is a *model* of a multi-adjoint normal logic program $\mathbb{P}$ if all weighted rules in $\mathbb{P}$ are satisfied by I.

Now, we recall the notion of stratified logic program, according to the seminal version of the notion, simultaneously introduced in [1, 17]. Given the set of propositional symbols appearing in the program $\mathbb{P}$, which will be denoted as $\Pi_{\mathbb{P}}$, a *stratification of* $\mathbb{P}$ is a mapping $||\cdot|| \colon \Pi_{\mathbb{P}} \to \mathbb{Z}^+$ such that, for each rule of $\mathbb{P}$ of the form $\langle p \leftarrow_i @\left[p_1, \ldots, p_m, \neg p_{m+1}, \ldots, \neg p_n\right]; \vartheta\rangle$, the following statements hold:

- $||p_i|| \leq ||p||$ for each $i \in \{1, \ldots, m\}$.
- $||p_i|| < ||p||$ for each $i \in \{m + 1, \ldots, n\}$.

Given a propositional symbol $p \in \Pi_{\mathbb{P}}$, the *stratum* of p is the integer $||p||$. The existence of stratifications characterizes the so-called stratified multi-adjoint normal logic programs. A multi-adjoint normal logic program $\mathbb{P}$ is *stratified* if there exists a stratification of $\mathbb{P}$. Indeed, if $||\cdot||$ is a stratification of $\mathbb{P}$, then the unique stable model of $\mathbb{P}$ can be computed in terms of a certain sequence of interpretations $\{M_1^{||\cdot||}, M_2^{||\cdot||}, \ldots\}$ related to the strata of $\mathbb{P}$ with respect to $||\cdot||$.

3 Normal Logic Programs Through Hypergraphs

Basic notions of (hyper)graph theory can be found in [2]. A *directed hypergraph* is a pair (V, E), where elements of E are *directed hyperedges* or *hyperarcs*, represented as ordered pairs $e = (T(e), H(e))$, with disjoint subsets of vertices $T(e)$ (the *tail*) and $H(e)$ (the *head*) [12]. Hereafter, directed hypergraphs are simply referred to as hypergraphs. An *edge labeling* is a function from E to labels, and a hypergraph with an edge labeling is an *edge-labeled hypergraph*. A *B-graph* is a hypergraph where all hyperarcs have only one element in their heads [12]. This paper focuses on labeled B-graphs, a natural representation of rules in a multi-adjoint normal logic program. This section will illustrate how a flexible MANLP can be represented by a specific edge-labeled directed B-graph.

Given a multi-adjoint normal logic program $\mathbb{P}$, we compute an associated B-graph $\mathcal{H}_{\mathbb{P}}$. Vertices correspond to propositional symbols in $\Pi_{\mathbb{P}}$, and hyperarcs are assigned for each rule. Labels include a 4-vector $(i, @, \{B_{s+1}, \dots, B_r\}, \vartheta)$, with special cases for single symbols in the body. Figure 1 shows the B-graph associated with Example 1. The resulting hypergraph is an edge-labeled directed B-graph, and the original program can be reconstructed from it [9, 10]. This characterization allows leveraging existing results and algorithms for directed hypergraphs in the analysis of fuzzy logic normal programs.

Given a directed hypergraph $\mathcal{H} = (V, E)$, two vertices $u, v \in V$, and a natural number q, a *path* in $\mathcal{H}$ from u to v of length q is a sequence of nodes and hyperarcs: $P_{uv} = \langle v_1, E_1, v_2, E_2, \dots, v_q, E_q, v_{q+1}\rangle$ where: $v_1 = u$, $v_{q+1} = v$, $u \in T(E_1)$, $v \in H(E_q)$, and $v_i \in H(E_{i-1}) \cap T(E_i)$, for all $i \in \{2, \dots, q\}$. The trivial sequence $P_{vv} = \langle v\rangle$, with $v \in V$, will be a path of length 0. Given a hypergraph $\mathcal{H}$, vertices u and v are *strongly path-connected* if there is a path from u to v and vice versa. A *strongly path-connected component* (SPC-component) is an equivalence class under this relation.

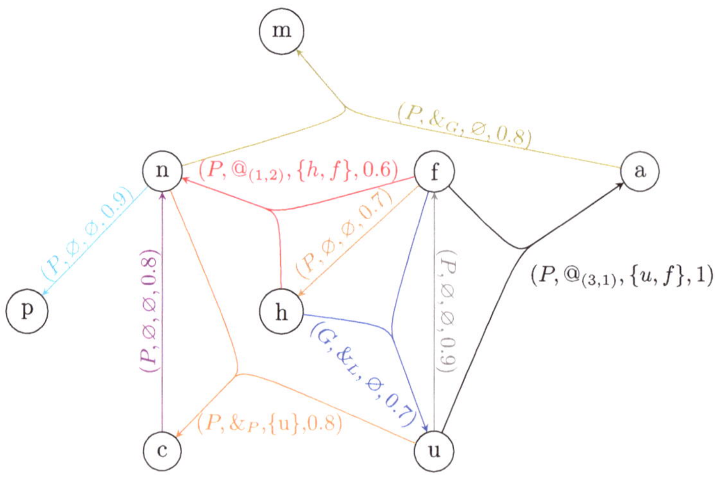

Fig. 1 Edge-labeled B-graph associated with the program given in Example 1. The facts of the program (rules 8, 9 and 10 of Example 1) are not shown by simplicity

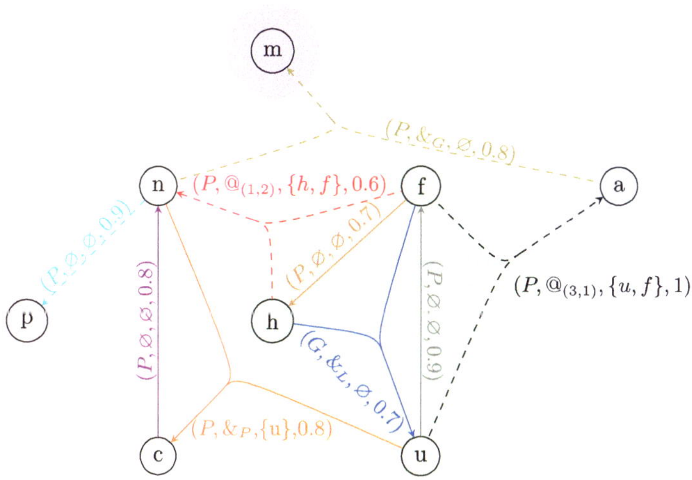

Fig. 2 SPC-components of the edge-labeled B-graph associated with the program given in Example 1. The d-hyperarcs are shown with dashed lines

Fig. 3 p-condensation digraph of the hypergraph in Fig. 1. Vertices are labelled with the vertices of the corresponding SPC of the hypergraph

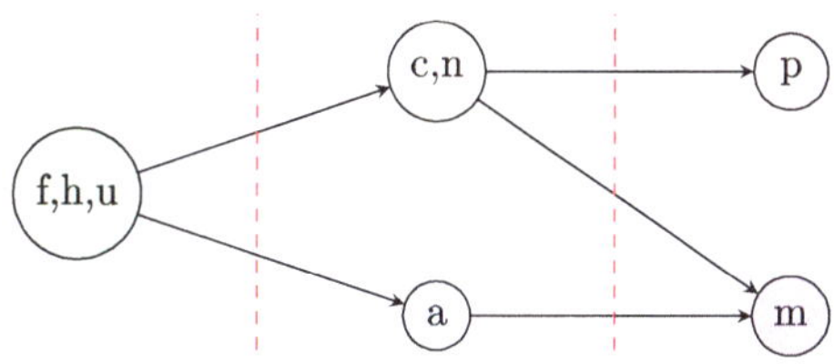

As a consequence, vertices in each cycle in a hypergraph belong to the same SPC-component. Example 1 has three SPC-components: {f, h, u}, {c, n}, and {a}.

SPC-components provide an interesting partition of the hyperarcs into two subsets. A *d-hyperarc* has vertices in **d**ifferent SPC-components in its tail and head; an *s-hyperarc* has at least one vertex in the **s**ame SPC-component. Figure 2 depicts the labeled B-graph associated with Example 1, highlighting SPC-components and d-/s-hyperarcs. Labels are also included. Notice that, the negation operators only appear in the d-hyperarcs.

The *path-condensation* (*p-condensation*, for short) of a hypergraph $\mathcal{H}$, denoted as $C(\mathcal{H})$, is a digraph whose vertices are SPC-components, with arcs connecting components based on hyperarc relationships [13]. Figure 3 shows the p-condensation of the hypergraph in Fig. 1.

4 Application of the Hypergraph Representation

This section shows that the hypergraph representation provides a mechanism for detecting whether a program $\mathbb{P}$ is stratified and, in this case, providing an "optimal" stratification.

As aforementioned, stratifications characterize the unique stable model of a stratified MANLP $\mathbb{P}$. However, since many stratifications can be defined from a program, even if we limit to non-isomorphic stratifications, then there might be multiple forms for computing the (unique) stable model of $\mathbb{P}$. For example, a stratification $||\cdot||$ with many strata requires many elements of the sequence $\{M_i^{||\cdot||}\}$, i.e. many steps, to obtain the unique stable model, thus slowing down its computation.

Furthermore, and more important, the constructive method of the stable model implies the computation of a succession of interpretations in a sequential way, avoiding parallel computations. The hypergraph representation of MANLP given in the previous section provides an interesting tool to deal with the two previous issues, and to optimize the computation of the unique stable model of stratified MANLP. Specifically, we obtain the following result.

Theorem 1 *Given a MANLP $\mathbb{P}$. If the labels of s-hyperarcs of $\mathcal{H}_{\mathbb{P}}$ do not contain negative variables in the same* SPC*-component of the s-hyperarc head, then the p-condensation graph $\mathcal{C}(\mathcal{H}_{\mathbb{P}})$ provides a stratification with the least possible number of non-empty strata, satisfying that $\|v_i\| < \|v_j\|$ for every couple of vertices v_i, v_j of $\mathcal{C}(\mathcal{H}_{\mathbb{P}})$ with a directed path from v_i to v_j.*

For example, in the previous particular program, we can see in Fig. 3 that the p-condensation graph offers an stratification $\|\cdot\|$ with three strata: $\|\mathrm{f}\| = \|\mathrm{h}\| = \|\mathrm{u}\| = 0$, $\|\mathrm{c}\| = \|\mathrm{n}\| = \|\mathrm{a}\| = 1$, and $\|\mathrm{p}\| = \|\mathrm{m}\| = 2$. Clearly, other stratifications can be defined on $\mathbb{P}$ satisfying that $\|v_i\| < \|v_j\|$ for every v_i, v_j of $\mathcal{C}(\mathcal{H}_{\mathbb{P}})$ with a directed path from v_i to v_j, but all of them have three or more strata.

Therefore, the stratification provided by the hypergraph representation of a MANLP $\mathbb{P}$ optimizes the number of strata. Furthermore, it also allows the computation of the least model of the subprogram associated with every stratum in parallel, with respect to each independent SPC-component of the stratum.

5 Conclusions and Future Work

We have complemented previous advances in order to completely characterize a MANLP through a hypergraph. We have also shown that, if a given MANLP is stratified, then the p-condensation graph of the hypergraph associated with the program provides a stratification with the least number of non-empty strata, satisfying that if two vertices of the graph are connected they must be in different strata, also optimizing parallel computations.

In the future, more properties and advantages will be obtained from the relationship between hypergraph theory and logic programming. For example, we will study the theoretical complexity of the computation of the stratification given by Theorem 1 and it will be compared with other traditional methods to compute stable models. Furthermore, we will apply the hypergraph representation to obtain new results in the computation of the semantics of modular logic programs [3].

Acknowledgements Partially supported by the 2014–2020 ERDF Operational Programme in collaboration with the State Research Agency (AEI) in projects PID2019-108991GB-I00 and PID2022-137620NB-I00, and the Ecological and Digital Transition Projects 2021 of the Ministry of Science and Innovation in project TED2021-129748B-I00.

References

1. Apt, K.R., Blair, H.A., Walker, A.: Towards a theory of declarative knowledge. In: Foundations of Deductive Databases and Logic Programming, pp. 89–148. Morgan Kaufmann (1988)
2. Berge, C.: Graphs and Hypergraphs. Elsevier Science Ltd. (1985)
3. Cabalar, P., Fandinno, J., Lierler, Y.: Modular answer set programming as a formal specification language. Theory Pract. Logic Program. **20**(5), 767–782 (2020)
4. Cornejo, M.E., Lobo, D., Medina, J.: Syntax and semantics of multi-adjoint normal logic programming. Fuzzy Sets Syst. **345**, 41–62 (2018)
5. Cornejo, M.E., Lobo, D., Medina, J.: Extended multi-adjoint logic programming. Fuzzy Sets Syst. **388**, 124–145 (2020)
6. Cornejo, M.E., Medina, J.: Impact Zadeh's theory to algebraic structures. Multi-adjoint algebras. J. Pure Appl. Math. **12**, 126–141 (2021)
7. Cornejo, M.E., Medina, J., Ramírez-Poussa, E.: Multi-adjoint algebras versus non-commutative residuated structures. Int. J. Approx. Reason. **66**, 119–138 (2015)
8. Díaz-Moreno, J.C., Medina, J., Portillo, J.R.: Towards the use of hypergraphs in multi-adjoint logic programming. Stud. Comp. Intell. **796**, 53–59 (2019)
9. Díaz-Moreno, J.C., Medina, J., Portillo, J.R.: Fuzzy logic programs as hypergraphs. Termination results. Fuzzy Sets Syst. **445**, 22–42 (2022)
10. Díaz-Moreno, J.C., Medina, J., Portillo, J.R.: Hypergraphs in logic programming. In: Bouraoui, Z., Vesic, S. (eds.) Symbolic and Quantitative Approaches to Reasoning with Uncertainty, pp. 442–452. Springer, Cham (2024)
11. Fitting, M.: The family of stable models. J. Logic Program. **17**(2–4), 197–225 (1993)
12. Gallo, G., Longo, G., Pallottino, S., Nguyen, S.: Directed hypergraphs and applications. Discrete Appl. Math. **42**(2–3), 177–201 (1993)
13. Harary, F., Norman, R.Z., Cartwright, D.: Structural Models: An Introduction to the Theory of Directed Graphs, 4th edn. Wiley, New York (1965)
14. Lloyd, J.: Foundations of Logic Programming. Springer (1987)
15. Madrid, N., Ojeda-Aciego, M.: On the existence and unicity of stable models in normal residuated logic programs. Int. J. Comput. Math. **89**(3), 310–324 (2012)
16. Medina, J., Ojeda-Aciego, M., Vojtáš, P.: Multi-adjoint logic programming with continuous semantics. Lecture Notes in Artificial Intelligence, vol. 2173, pp. 351–364
17. Van Gelder, A.: Negation as failure using tight derivations for general logic programs. J. Logic Program. **6**(1–2), 109–133 (1989)
18. Zhang, C., Chen, L., Zhao, Y.-P., Wang, Y., Chen, C.L.P.: Graph enhanced fuzzy clustering for categorical data using a Bayesian dissimilarity measure. IEEE Trans. Fuzzy Syst. **31**(3), 810–824 (2023)

Enhancement of Discrete Bacterial Memetic Evolutionary Algorithm for Solving the Travelling Repairman Problem

Ali Jawad Ibada, Boldizsár Tüű-Szabó, and László T. Kóczy

Abstract The Traveling Repairman Problem (TRP) is concerned with repairing a set of locations rather than visiting them. In this paper, we propose an enhanced version of the Discrete Bacterial Memetic Evolutionary Algorithm (DBMEA) to solve TRP. DBMEA is combining with a new method for generating the initial individual candidate solution which is called Circle Group Heuristic (CGH). CGH is constructed with the help of Genetic Algorithm (GA). The enhanced version of DBMEA with CGH has been tested for several benchmark reference data of TRP. The results show that the enhanced version has a faster and better solutions for most cases in comparison to state-of-the-art heuristics mentioned in the literature. Furthermore, for larger benchmark instances, it provided better solutions than the previously best-known results. These test results support the claim that the DBMEA with CGH is the most effective approach and recommend its use for the Traveling Repairman Problem, particularly for large instances.

Keywords TRP · CGH · DBMEA · Memetic algorithm

1 Introduction

In TRP, a repairman needs to visit a set of locations to perform repairs. The objective is to Minimize the total travel times at each node of a repairman who is responsible for maintaining a set of locations, where each location has a different frequency of maintenance needs. The TRP arises in a variety of real-world applications, such as

A. J. Ibada · L. T. Kóczy
Department of Telecommunications and Media Informatics, Budapest University of Technology and Economics, Budapest, Hungary
e-mail: ali.alshukri@tmit.bme.hu

B. Tüű-Szabó (✉) · L. T. Kóczy
Department of Logistics, Széchenyi István University, Gyor, Hungary
e-mail: tuu.szabo.boldizsar@ga.sze.hu

L. T. Kóczy et al. (eds.), *Computational Intelligence and Mathematics for Tackling Complex Problems 6*, Studies in Computational Intelligence 1222,
https://doi.org/10.1007/978-3-031-97879-1_18

maintenance scheduling for equipment or vehicles, and can be used to minimize travel time and costs while maximizing productivity.

TRP can be define as a graph search problem in a weighted edge as showing in Eq. 1.

$$G_{TRP} = (V_{cities}, E_{conn}) \tag{1a}$$

$$V_{cities} = v_0 \mathrm{U}\{v_1, v_2, \ldots, v_n\} \tag{1b}$$

$$E_{conn} \subseteq \left\{(V_i, v_j) | i \neq j\right\} \tag{1c}$$

$$C : V_{cities} \times V_{cities} \rightarrow R, C = (c_{ij})_{(n+1)\times(n+1)} \tag{1d}$$

C is called cost matrix where c_{ij} is the cost of going from vertex i to vertex j. The goal is to find a permutation of the vertices ($p_0, p_1, p_2, p_3, \ldots, p_n$) that minimalizes the sum of arrival times (C_{sum}) at all nodes (Eq. 2).

Minimize the sum of arrival times:

$$\mathrm{C}_{\mathrm{p_i}} = \sum_{\mathrm{j=1}}^{\mathrm{i}} \mathrm{C}_{\mathrm{p_{j-1}}}, \mathrm{P_j}$$

where the arrival time at node $\mathrm{p_i}$ is:

$$\mathrm{C_{sum}} = \sum_{\mathrm{i=1}}^{\mathrm{n}} \mathrm{C}_{\mathrm{p_i}} \tag{2}$$

Since TRP belongs to the family of the Traveling Salesman Problem (TSP), TRP is a NP-hard problem [1]. In 2011 Salehipour et al. proposed the first metaheuristics algorithm to solve TRP [2]. The algorithm consists of two stages, a tour construction stage, and an improvement stage. GILS-RVND heuristic is the most efficient heuristic found in the literature which developed by Silva et al. in 2012 [3]. It is a two-phase method. The initial tours which are constructed using GRASP, and the improvement phase which consist of Iterated Local Search (ILS) and Variable Neighborhood Descent with Random Neighborhood Ordering (RVND). An effective memetic algorithm for the generalization of the TRP is proposed by Ngueveu et al. They added capacity restrictions and a homogeneous fleet of vehicles to the Cumulative Capacitated Vehicle Routing Problem (CCVRP) for investigation [4]. Ban in 2017 presents a metaheuristic algorithm combines the Variable Neighborhood Search and Tabu Search [5]. Another metaheuristic method with two phases was also proposed by Ramadhan et al. in 2019, it was only tested only on very small instances [6]. A multi-level parallelization approach was proposed by Araujo et al. in

2018 for solving the TRP in fast execution time, but it has lower accuracy compared to the state-of-the-art methods [7].

We presented the efficient DBMEA in 2016, it was used successfully to solve the TSP [8]. Then, the DBMEA was improved with certain speed-up techniques in the local search, which led to a significant improvement of runtime, while keeping the same quality of results for TSP instances [9]. Furthermore, our experimentation on the Traveling Salesman Problem (TSP) using CGH with DBMEA, it yielded superior solutions with reduced runtime [10, 11].

2 The Discrete Bacterial Memetic Evolutionary Algorithm

DBMEA represents a memetic algorithm, integrating the highly effective bacterial evolutionary algorithm for global optimization with a simple combinatorial local search [8]. This fusion overcomes the drawbacks associated with each method. The evolutionary algorithms explore the global search space, their practical utility often results in a quasi-optimal solution due to their relatively slow convergence speed. While the local search methods focus on a specific neighborhood around the current solution with rapidly converging to the nearest local optimum [12].

The DBMEA has four stages: initial population creation, bacterial mutation (coherent segment mutation and loose segment mutation), local search (2-OPT and 3-OPT), and gene transfer which is cyclically repeated.

2.1 *The Initial Population Stage*

The individuals are randomly generated during the initial population stage. Each individual signifies a possible solution which representing the potential tours of the TRP.

2.2 *Bacterial Mutation Stage*

During this stage, each bacterium within the population undergoes individual treatment. Firstly, A specific number of clones is generated from the original bacterium, and this number is predefined as (N_{clones}). Secondly, the original bacterium and its clones undergo subdivision into chromosomes, each possessing a fixed length denoted as (I_{seg}). Thirdly, the algorithm employs two types of mutation: coherent segment mutation and loose segment mutation. Figure 1 shows coherent segment mutation.

Next, the segments of the bacterium are randomly modified in the clones, while the original bacterium remains unaltered. Then, the fitness values of the clone bacteria,

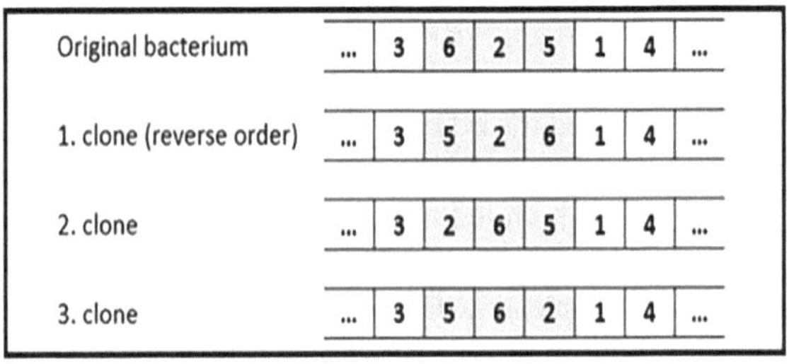

Fig. 1 Coherent segment mutation

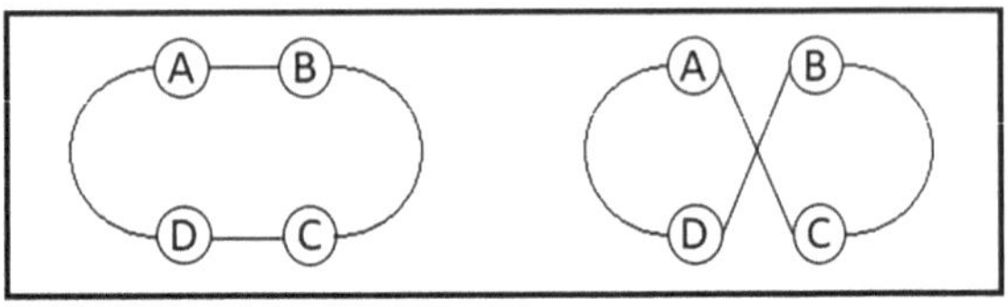

Fig. 2 2-OPT local search

including the original, are evaluated. If any clone has better fitness than the original bacterium, the segment from the better clone is transferred back to both the original bacterium and all other clones.

2.3 Local Search Stage

Individual improvement is pursued in this stage. The method involves a thorough examination of reorganizing DBMEA, using 2-opt and 3-opt methods which enhance individuals by optimizing bounded size sub-graphs, focusing on local optimization.

2.3.1 2-Opt Local Search

The 2-opt local search operates by replacing two edge pairs in the original graph with the aim of reducing the length of the tour. The iterative process involves replacing two edge pairs (AB, CD) with AC and BD edges, resulting in a newly constructed possible tour. If this tour demonstrates a lower cost than the original, the edge pairs AB and CD are exchanged. Figure 2 shows 2-OPT local search.

2.3.2 3-Opt Local Search

The 3-opt local search enhances the tour by substituting three edges with three alternatives. Figure 3 illustrates the potential replacements in the 3-opt local search. This 3-opt stage local search is the most time-consuming during execution. The local

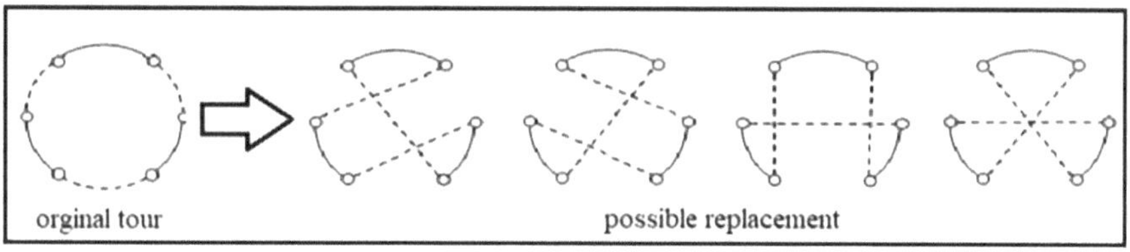

Fig. 3 3-OPT local search

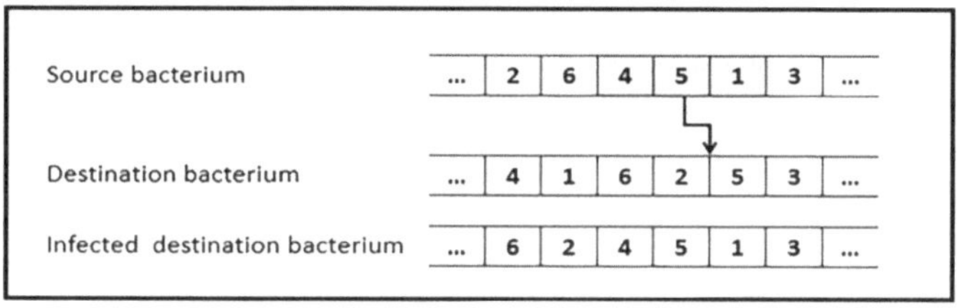

Fig. 4 Gene transfer

search concludes when the tour cannot be further improved through additional 3-opt steps.

2.4 *Gene Transfer Stage*

The population is initially arranged in decreasing order based on their fitness values in this phase, which are then sorted and divided into two halves (a superior and an inferior half). The following process repeated N_{inf} times: one random bacterium is selected (source bacterium) from the superior half and another random bacterium (destination bacterium) from the inferior half. A randomly selected segments with length ($I_{transfer}$) is transferred from the source bacterium to the destination bacterium. Figure 4 illustrates the segment transfer in the gene transfer stage.

3 Circle Group Heuristic

We have introduced a novel approach that quasi-optimizes, ensuring definite improvement of the initial population for the TRP. which will be called Circle Group Heuristic (CGH). CGH applying the concept of “Neighborhood Circle” (NC). The NC has a pre-specified radius which will limit and speed up the search for the best possible initial population candidate. Initially, the starting node, City 0, serves as the center of a circle, and an invisible circle is drawn with a predefined radius (R). Subsequently, the nearest unvisited city within the circle is visited. Once all the cities inside

Fig. 5 The procedure of creating CGH

1. Center = city 0
2. Create invisible circle with radius (R)
3. Visit all cities inside the circle in NN mode
4. If (unvisited cities outside the circle are exist)
   - Center = nearest city outside the circle
   - Go to step 2
5. Else
   - Next city = city 0

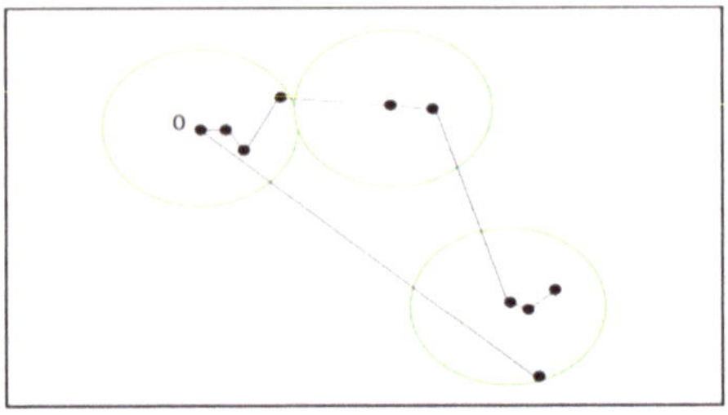

Fig. 6 Example for a simple CGH tour

the circle have been visited, the tour proceeds to the nearest unvisited city outside the circle. This newly visited city outside the circle becomes the center of a new circle, and the process repeats until all cities have been visited. Figure 5 illustrates the procedure for creating CGH, while Fig. 6 shows a simple tour generated by the CGH.

The most effective radius of the circle in CGH, which yields superior tours, varies from one graph to another. Through an examination of various countries with their corresponding city and road networks as benchmarks, it was observed that countries differ from each other, likely due to their distinct topologies. The optimal radius of the circle may depend on factors such as the total area of the country, the number of cities, and, notably, the distribution of cities within the country.

We tested two methods to calculate R, Brute force optimization and Genetic Algorithm (GA). We found that GA gives better radius for CGH within reasonable time.

4 Results

DBMEA with CGH initial population was tested on TSPLIB (TSP Library) benchmark instances [13], with each instance being executed 20 times and taking the average value of them. The obtained results were then compared with the outcomes of previous version DBMEA, which is already recognized as superior to state-of-the-art heuristics [12]. Across all instances, DBMEA with CGH demonstrated faster execution times in comparison to normal DBMEA. Table 1 shows the small instances

Table 1 Solutions for small instances

Instance	Average time DBMEA		Average time DBMEA with CGH			
			R (GA)	S20	R (GA)	S50
	S20	S50				
TRP-Sn-R1	0.46620	5.3862	35.266	**0.46435**	90.421	**4.504**
TRP-Sn-R2	1.81035	4.157	63.565	**1.20820**	92.945	**2.7564**
TRP-Sn-R3	0.31860	5.8018	30.802	**0.26175**	6.839	**4.9896**
TRP-Sn-R4	0.64560	10.2186	7.974	**0.58460**	7.165	**9.6592**
TRP-Sn-R5	0.86665	7.1022	17.916	**0.69775**	6.836	**6.4844**
TRP-Sn-R6	0.36460	33.613	1	**0.29005**	10.840	**33.01**
TRP-Sn-R7	0.81975	3.9832	7.591	**0.78280**	4.503	**3.2212**
TRP-Sn-R8	0.73525	10.0588	29.723	**0.72810**	1	**8.2512**
TRP-Sn-R9	0.62660	9.1434	61.948	**0.60420**	8.567	**8.1272**
TRP-Sn-R10	0.66405	7.851	96.939	**0.60135**	5.202	**6.7018**

Bold values indicate cases where the new DBMEA with CGH outperformed the original DBMEA, either through reduced execution time or improved tour quality

(20 and 50) which are generated by Salehipour et al. The table observes a reduction in execution time by an average of 20.45% and 14.29 for instances 20, 50 respectively compared with the previous version of DBMEA. Real cities are shown in Table 2 with large instances (up to 226 nodes). The new DBMEA with CGH shows a reduction in execution time by an average of 8.69% compared with DBMEA as well as it gives better tour value for *rat195* and *pr226*.

Table 2 DBMEA and DBMEA with CGH results for the TSPLIB instances

Instance	Best known	DBMEA			DBMEA with CGH				
		Best value	Avg. value	Avg. sec	R (GA)	Initial CGH	Best value	Avg. value	Avg. sec
pr76	3,455,242	3,455,242	3,455,242	8.1218	1	3,696,829.0722	3,455,242	3,455,242	**7.3098**
rat99	57,986	57,986	57,986	54.6843	6.093	62,967.779	57,986	57,986	**51.8731**
kroA100	983,128	983,128	983,128	24.0905	74.407	1,103,528.624	983,128	983,128	**21.7378**
kroD100	976,965	976,965	976,965	19.5177	137.514	1,054,638.602	976,965	976,965	**18.2528**
lin105	603,910	603,910	603,910	57.0368	51.203	687,043.793	603,910	603,910	**47.8758**
pr107	2,026,626	2,026,626	2,026,626	8.23	367.020	2,083,979.961	2,026,626	2,026,626	**7.4696**
rat195	210,191	218,665	218,693.1	157.8004	1	228,467.179	218,665	**218,665**	**143.2586**
pr226	7,100,308	7,196,869	7,196,898.6	208.2368	302.981	10,785,832.078	7,196,869	**7,196,869**	**200.74275**

Bold values indicate cases where the new DBMEA with CGH outperformed the original DBMEA, either through reduced execution time or improved tour quality

5 Conclusion

In this paper, we introduced a novel and efficient tour construction method to improve DBMEA for solving the TRP. Based on the computational res can be asserted that better initial individuals contribute to a faster convergence speed. Based on our experimental findings, we recommend incorporating our newly proposed Circle Group Heuristic (CGH) in the initial population creation stage of the DBMEA. Furthermore, it is highly likely that CGH could prove beneficial for other heuristic optimization algorithms aimed at solving the Traveling Salesman Problem (TRP).

References

1. Aho, A.V., Hopcroft, J.E.: The Design and Analysis of Computer Algorithms. Pearson Education India (1974)
2. Salehipour, A., Sörensen, K., Goos, P., Bräysy, O.: Efficient GRASP+VND and GRASP+VNS metaheuristics for the traveling repairman problem. 4OR Q. J Oper. Res. **9**(2), 189–209 (2011)
3. Silva, M.M., Subramanian, A., Vidal, T., Ochi, L.S.: A simple and effective metaheuristic for the Minimum Latency Problem. Eur. J. Oper. Res. **221**(3) (2012)
4. Ngueveu, S., Prins, C., Calvo, R.W.: An effective memetic algorithm for the cumulative capacitated vehicle routing problem. Comput. Oper. Res. **37**(11) (2010)
5. Ban, H.B., Nguyen, D.N.: A meta-heuristic algorithm combining between Tabu and variable neighborhood search for the minimum latency problem (2017)
6. Ramadhan, F., Imran, A.: A two-phase metaheuristic method for solving travelling repairman problem. In: 2019 International Conference on Sustainable Engineering and Creative Computing (ICSECC), Bandung, Indonesia, pp. 155–159 (2019)
7. Pereira Araujo, R., Coelho, I.M., Marzulo, L.A.J.: A DVND local search implemented on a dataflow architecture for the minimum latency problem. In: 2018 IEEE International Parallel and Distributed Processing Symposium Workshops (IPDPSW), Vancouver, BC, Canada, pp. 1250–1259 (2018)
8. Kóczy, L.T., Földesi, P., Tüű-Szabó, B.: A discrete bacterial memetic evolutionary algorithm for the traveling salesman problem. In: IEEE World Congress on Computational Intelligence (WCCI 2016), Vancouver, Canada, pp. 3261–3267 (2016)
9. Tüű-Szabó, B., Földesi, P., Kóczy, L.T.: Improved discrete bacterial memetic evolutionary algorithm for the traveling salesman problem. In: Proceedings of the Computational Intelligence in Information Systems Conference (CIIS 2016), Bandar Seri Begawan, Brunei, pp. 27–38 (2017)
10. Ibada, A.J., Tüu-Szabó, B., Kóczy, L.T.: A new efficient tour construction heuristic for the Traveling Salesman Problem. In: Proceedings of the 2021 5th International Conference on Intelligent Systems, Metaheuristics & Swarm Intelligence (ISMSI '21), pp. 71–76. Association for Computing Machinery, New York, NY, USA (2021)
11. Ibada, A.J., Tüu-Szabó, B., Kóczy, L.T.: Effect of the initial population construction on the DBMEA algorithm searching for the optimal solution of the traveling salesman problem. Infocommun. J. **XIV**(3), 72–78 (2022)
12. Tüu-Szabó, B., Földesi, P., Kóczy, L.T.: An efficient evolutionary metaheuristic for the traveling repairman (minimum latency) problem. Int. J. Comput. Intell. Syst. **13**, 781–793 (2020)
13. http://comopt.ifi.uni-heidelberg.de/software/TSPLIB95/

Simulated Annealing and Bacterial Foraging for Probabilistic Neural Network Parameters Adjustment

Szymon Kucharczyk and Piotr A. Kowalski

Abstract Probabilistic Neural Networks (PNNs), a category of Feedforward Neural Networks, leverage Kernel Density Estimators (KDEs) and the Bayesian conditional probability theorem for estimating conditional probabilities. Initially designed for classification, these networks exhibit commendable performance in both classification and regression tasks. The training process involves determining optimal or suboptimal values for the KDE smoothing parameter, commonly accomplished through analytical methods such as the Plug-in technique. Additionally, metaheuristic approaches like Particle Swarm Optimisation and Krill Herd Algorithm have been employed for smoothing parameter optimisation in PNNs due to the absence of gradient calculations. This contribution proposes the integration of Bacterial Foraging Optimisation (BFO) and Simulated Annealing (SA) for enhancing PNNs. The efficiency of these techniques in optimising PNNs is compared with the conventional Plug-in method, employing benchmark classification datasets sourced from UCI and Kaggle repositories. The results reveal that SA surpasses other methods in specific benchmarking tasks, suggesting its efficacy in training PNNs for specific problem domains.

Keywords Probabilistic neural networks · Metaheuristics · Simulated annealing · Bacterial foraging optimisation

S. Kucharczyk
AGH Doctoral School, AGH University of Krakow, Krakow, Poland
e-mail: kucharcz@agh.edu.pl

P. A. Kowalski (✉)
Faculty of Physics and Applied Computer Science, AGH University of Krakow, Kraków, Poland
e-mail: pkowal@agh.edu.pl

Systems Research Institute, Polish Academy of Sciences, Warsaw, Poland

L. T. Kóczy et al. (eds.), *Computational Intelligence and Mathematics for Tackling Complex Problems 6*, Studies in Computational Intelligence 1222,
https://doi.org/10.1007/978-3-031-97879-1_19

1 Introduction

Memory-based networks known as Probabilistic Neural Networks were introduced for classification purposes by Specht in [15], wherein the model is directly constructed from the training data samples. PNNs are particularly effective when the training data serves as a good representation of a target data distribution and undergoes iterative updates with new samples, PNNs operate by estimating class probabilities through the utilization of a Bayesian conditional probability theorem. This approach facilitates the incorporation of prior information from the data into the network structure.

As elucidated earlier, the weights of PNNs are constructed using both training samples and Kernel Density Estimators (KDE). The sole trainable parameter in a basic Probabilistic Neural Network is the KDE smoothing parameter denoted as h. The optimisation of PNN parameters can be achieved through various approaches, including statistical methods such as the Plug-in technique [8] and cross-validation (CV) [6], as well as computational intelligence techniques like PSO [2], BAT-algorithm [10], or Reinforcement Learning [9].

Recent advancements in PNN training involve the successful application of various metaheuristics, including PSO [2, 6], BAT-algorithm [10], FPA [7], and ABC [14]. The superiority of nature-inspired methods over traditional ones lies in their adaptability for optimising diverse tasks and ensuring fair randomization of results [10]. Moreover, they exhibit efficiency in conducting a thorough search on both global and local scales within a reasonable timeframe [7].

This contribution presents an original performance comparison of PNN training optimisation between the Plug-in technique and two metaheuristics: Simulated Annealing (SA) [5] and Bacterial Foraging Optimisation (BFO) [11]. These computational intelligence methods, not previously utilized for PNN training, may signify a progression over existing techniques.

The paper is organized as follows: first, a theoretical introduction to PNNs and their training with metaheuristics is provided in 2. Subsequently, PNN optimisation techniques are expounded upon in Sect. 3. Following that, the data and validation approach are detailed in Sect. 4. The results are then presented in Sect. 5, and the paper is summarized in Sect. 6.

2 Probabilistic Neural Networks

Probabilistic Neural Networks, a subtype of feedforward neural networks, are constructed based on Parzen window estimation utilizing Kernel Density Estimation and Bayesian conditional probability (P_b) estimation for determining output class (refer to [7, 15]). The KDE employed for P_b estimation at the summation layer of PNN was initially introduced in [15]:

$$\hat{f}(\mathbf{x}) = \frac{1}{mh_n} \sum_{i=1}^{m} K\left(\frac{\mathbf{x} - \mathbf{x}_i}{h}\right), \tag{1}$$

where m represents the training data cardinality, n denotes the dimensionality of $\mathbf{x}$, $\mathbf{x}_i$ stands for the i-th training pattern, K denotes a kernel function and finally h represents Kernel's smoothing parameter.

Various kernel functions can be employed to construct PNNs. In this study, the Gaussian kernel function was chosen due to its well-established nature in the research field and ease of interpretability, allowing a focused comparison of training methods [8, 15]. The kernel function for the sample vector $\mathbf{x}$ is defined as follows:

$$K(x) = \frac{1}{\sqrt{2\pi}} e^{\frac{-\mathbf{x}^T\mathbf{x}}{2}}. \tag{2}$$

In the context of this study, the convention of product kernels was adopted, necessitating the separate calculation of $K(x)$ for each data feature [6].

Originally, PNNs are constructed with four layers (refer to [10, 15]): the *Input layer*, which takes the input vector; the *Pattern layer*, with the number of neurons equal to the number of training samples and responsible for KDE; the *Summation layer*, comprising one neuron for each class, aggregating results from the previous layer; and the *Output layer*, which contains a single neuron calculating predictions using Bayes' theorem.

The training procedure for PNNs involves the determination of the optimal or suboptimal smoothing parameter, denoted as h. Initially, this process was executed using a brute-force method [15] and analytical methods such as Approximation or Plug-in [6, 8]. The approach selection is typically based on the characteristics of the data and the type of kernel used (radial or product kernels). To enhance PNN efficiency across diverse problem domains, computational intelligence techniques have been applied [6, 7, 10]. Metaheuristic algorithms present an advancement over traditional methods as they perform gradient-free, randomised global and local optimisation without constraints [6]. Constraints arising from the approximation in statistical methods often limit their convergence. The flexibility of metaheuristics addresses exploration/exploitation issues, facilitating the identification of the optimal h parameter value within a reasonable timeframe [7]. Moreover, nature-inspired techniques are simple to implement, requiring only a straightforward cost function and parameter adjustment.

3 Training Methods

3.1 Plug-In

This technique determines the smoothing parameter based on statistical equations by estimating Mean Integrated Square Error (MISE) [8]. The h parameter value can be found by solving the following:

$$h_0 = \left(\frac{nW(K)}{U(K)^2 Z(f) m} \right)^{\frac{1}{(4d+1)}}, \tag{3}$$

where $d \geq 2$, $W(K)$ and $U(K)$ stand for the kernel function and $Z(f)$ represents the estimated density of the f function defined similarly to Kowalski et al. [6]. The value of the function $Z(f)$ is iteratively calculated using Eq. (3). In this study, the d parameter was set to *2* [8].

The optimisation approach employed here proves to be computationally efficient for product kernels. This is particularly relevant as the Plug-in method is solely established for one-dimensional data, necessitating the separate calculation of the smoothing parameter for each feature [6]. The resulting h value, computed through this technique, serves as an approximation to the true analytical value. Importantly, it can be swiftly determined when compared to alternative methods [6].

3.2 Bacterial Foraging Optimisation

BFO, proposed by K. Passino in 2022 [11], is a nature-inspired optimisation technique, motived by the foraging behavior of E. coli bacteria, living in human intestines. Normally, the bacterium performs its motions by flagella rotation, which enables it to swim, run, or tumble in the environment [11]. In nature, the bacterium seeks nutrition by constant variation of swimming and tumbling in random directions, which lets it find optimal food sources [1].

The BFO algorithms, simulate the E. coli behavior, consisting of several iterative steps, executed till the stopping condition is met. Similarly to other metaheuristics used for optimisation [7], it requires a bacterium population p, a definition of a bacterium as a solution vector Θ, and a fitness function $J(\Theta)$ as a representation of the nutrition quality. The optimisation process, after population initialization and initial cost function (error value) computation, is performed as follows [11]:

1. **Chemotactic**
   Normally, the bacterium motion is induced by the presence of chemical attractants (chemotaxes). In the chemotactic phase, this attraction makes E. coli move to the nearest food source. Usually, it is executed in multiple steps, for Nc iterations:

(a) *Tumbling*
This stage is responsible for the bacterium's random movement, described with the equation:

$$\Theta^i = \Theta^i + C(i) \times \frac{\Delta(i)}{\sqrt{\Delta^T(i)\Delta(i)}} \tag{4}$$

where i stands for the ith chemotactic step, $C(i)$ is the length of bacterial random walk (step size) and Δ represents a random vector $\in [-1, 1]$.

(b) *Swimming*
This step is responsible for the exploitation and in its principle is similar to tumbling. If the error value is lower after tumbling than it was before, the bacterium will continue swimming in the same direction for Ns times. However, if the error value after the tumbling or swimming step is higher than it was before, it will discard the solution and come back to random generation defined in 1a

(c) *Swarming*
It mirrors the group attraction mechanism in the bacteria foraging process. Basically, it adds a small error signal connected with a gradient generated by the group consuming the nutrient. It is steered with several parameters: $d_{attract}$, $\omega_{attract}$, $h_{repellant}$ and $\omega_{repellant}$. This step equation definition can be found in [11].

2. **Reproduction**
The bacteria population p is sorted by the ascending cost function value. Half of the population with the highest fitness function value dies and it is removed from p. To keep the p of constant shape, the remaining part of the bacterium is doubled and put into places of the removed bacterium.
3. **Elimination and Dispersal**
In nature, it is possible that some bacteria might be dispersed to new locations or eliminated, which is similar to the mutation process in Genetic Algorithms [1]. This stage presents an implementation of this behavior. First, a random vector $t \in [-1, 1]$ of a shape equal to the solution vector shape is generated. Then, random values of randomly selected bacterium vector representations are replaced with corresponding values of t. This stage is executed for N_{ed} steps and the elimination-dispersal rate is controlled with p_{ed} parameter.

The BFO procedure was applied to solve classical problems, like the traveling salesman problem, and Neural Network optimisation. Values of the BFO parameters used in this work are described in Sect. 4.

3.3 Simulated Annealing

Metaheuristics for the optimisation process are inspired by swarms, animals, music, and physical processes. The SA algorithm, introduced by Kirkpatrick et al. in 1983, is a physics-based optimisation process founded on the annealing phenomenon while growing crystals from melt [5]. In this phenomenon, a substance (e.g. crystal) is melted and then its temperature is slowly lowered in a controlled manner. It is an important process, as low temperature is not a sufficient condition for finding ground states of matter and the annealing process is required [5].

The key component of this procedure is the Boltzmann distribution, which quantifies the annealing process and originally is defined as:

$$p(x) = e^{\frac{-\Delta f(x)}{kT}} \tag{5}$$

where k stands for the Boltzmann constant, T represents the temperature and $f(x)$ is the configuration energy of variable x.

The SA requires a particle population p, a definition of a particle as a solution vector Φ, a maximum number of iterations over algorithm N, and a fitness function $J(\Phi)$ as a representation of the melting (food, nutrition for other metaheuristics) quality. The algorithm involves multiple i steps [13]:

1. Generate the initial population of particles p_0, calculate the cost function value for p_0 and initialize the temperature T.
2. For each particle, select randomly a new solution Φ_i in the particle neighbourhood using the equation

$$\Phi_i = \Phi_{i-1} + d\frac{N - i + 1}{N}\Delta_i \tag{6}$$

   where i is the iteration number, Δ is a random vector in range $\Delta \in [-2, 5]$ and d is a constant equal to 0.01. Then the fitness value $J(\Phi_i)$ is calculated, and the new position and cost value are saved.
3. If $J(\Phi_i) < J(\Phi_{i-1})$ new solution replaces the previous one as it has a lower value of the cost function. Otherwise, the Φ_i also might be accepted to be the new particle position with probability based on 5, where $\Delta f(x)$ is defined as $J(\Phi_i) - J(\Phi_{i-1})$ and k is substituted with 1.
4. Decrease the temperature gradually in each iteration:

$$T_i = T_{i-1} * \alpha^i \tag{7}$$

   where α is a float parameter (originally equal to 0.8 [5]).
5. Steps 2 to 4 are repeated till the stopping condition is met—either the maximum iteration number or minimal is met or the cost function minimum is achieved.

The SA method was applied to solve various tasks, including the physical design of Computers [5] and Convolution Neural Network structure optimisation [13]. Values of the algorithm parameters used in this research are presented in Sect. 4.

4 Data and Methods

In this study, various datasets were employed for the comparison of training methods, allowing the assessment of their performance across tasks with diverse characteristics. The majority of these datasets were sourced from the UCI repository [3], with the exception of the Ghouls-Goblins-and-Ghosts dataset, which was obtained from Kaggle [4]. The Iris dataset served as a benchmark due to its well-established status in the literature. Notably, the evaluation encompassed datasets with more than two classes, such as Iris, Wine, and Ghouls-Goblins-and-Ghosts. Datasets from the medical domain, namely the Indian Liver Patient Dataset (ILPD) and Breast Cancer, were also included. The Breast Cancer data, in particular, facilitates performance testing on data with a high level of dimensionality. To ensure the reproducibility of experiments, all computations were carried out with the NumPy random seed value set to 33. Additionally, datasets were preprocessed to remove 'NaN' values.

For this research, a stratified 5-fold cross-validation with 20% of validation set size was used to evaluate PNNs, in order to ensure reasonable model generalization [12]. To quantify the comparison, accuracy and categorical cross-entropy (log-loss) were employed [12]. Train scores, test scores, and fit time were calculated as a mean value from all folds. Finally, the fitness (cost) function was defined as *error rate* following [9]:

$$error\ rate = 1 - \frac{number\ of\ correct\ predictions}{cardinality\ of\ test\ sample}. \tag{8}$$

The PNNs have one trainable parameter—smoothing parameter h. Here, we are comparing four different levels of smoothing parameter optimisation following [7]: same value for whole data h, separate value for each attribute h_a, separate value for each class h_c and separate value for each class and attribute h_{ac}. To meet the requirements of the Plug-in technique, for this last case the smoothing parameter was calculated separately for each feature even for the h_c level. The smoothing parameter was initialized as a random vector of range $h_{init} \in [0, 10]$ [7]. In addition, smoothing values were limited to be greater than 0 during the optimisation process with BFO and SA, as required by the PNN architecture.

For BFO and SA techniques, the optimised solution (bacterium, particle) was defined as a vector with a shape respective to the selected smoothing parameter level. Indeed, a population is defined as a set of individuals, each of them containing all smoothing parameter values. These individuals represent a group of solutions to the PNN training problem. Here, the parameters of metaheuristic algorithms were determined using limited GridSearch. After an initial run of these methods on all datasets using default parameters [5, 11], the ILPD was selected for optimisation of the BFO algorithm, and the wine dataset was used for the GridSearch of the SA algorithm. Then, a maximum number of iterations exploration was performed for each method, with values [50, 100, 200, 300]. It showed that after 200 iterations, algorithms did not improve over the next iterations. After that, other parameters of nature-based methods were refined, using 200 as the maximum number of iterations N. For the BFO method, the following values were examined: C_i: [0.05, 0.1, 0.2], N_c: [4, 6, 8]. Similarly, for the SA technique, α: [0.4, 0.6, 0.8], T: [40, 100, 200, 500] were tested. The parameters selection for GridSearch was founded on the suggestion of parameter significance [5, 11]. In the end, the following parameters were used for all datasets to obtain final results:

- BFO:
  - *max number of iterations* = 200;
  - *population size* = 20;
  - $N_{ed} = 2$, $N_c = 4$, $N_s = 4$;
  - $C(i) = 0.1$;
  - $p_{ed} = 0.25$;
  - $d_{attract} = 0.1$, $\omega_{attract} = 0.2$;
  - $h_{repellant} = 0.1$, $\omega_{repellant} = 10$.
- SA:
  - *max number of iterations* = 200;
  - *population size* = 20;
  - *stopping temperature* = 1e−8;
  - $T = 100$;
  - $\alpha = 0.8$.

5 Results

Within this research, we examined SA and BFO optimisation techniques for determining optimal values of the PNN smoothing parameter for benchmark datasets. All techniques and parameters were described in Sect. 4 to ensure the experiment's reproducibility.

The comparison results are shown in Tables 1 , 2 , 3 , 4 and 5. For the Iris dataset, the analytical procedure had the highest cross-validation test accuracy score, although the BFO also performed reasonably well. In Table 2, outcomes of the examination for the Ghosts dataset are demonstrated. Similarly, to the Iris dataset Plug-in approach gave the best results, but here the BFO did not achieve satisfying metric scores. Indeed, for the Breast Cancer and Wine datasets (shown in Tables 3 and 4 respectively), nature-based techniques (BFO and SA), significantly outperformed the classical PNN

Table 1 Training methods comparison for the Iris dataset

	accuracy train/test				log_loss train/test				fit_time			
h type	h_c	h_ac	h	h_a	h_c	h_ac	h	h_a	h_c	h_ac	h	h_a
Plug-in	1.000/0.953	1.000/0.953	**1.000/0.967**	**1.000/0.967**	0.001/0.560	0.001/0.560	0.002/0.243	0.002/0.243	0.185	0.179	0.058	0.059
BFO	0.780/0.787	0.825/0.800	**0.988/0.967**	0.963/0.940	0.687/0.523	0.452/0.465	**0.082/0.141**	0.623/0.621	11.85	10.89	4.266	8.734
SA	1.000/0.527	0.987/0.740	1.000/0.693	0.990/0.773	0.185/12.01	0.261/6.266	0.004/0.768	0.089/2.524	0.203	1.400	0.039	1.000

Table 2 Training methods comparison for the Ghosts, Goblins, and Ghouls dataset

	accuracy train/test				log_loss train/test				fit_time			
h type	h_c	h_ac	h	h_a	h_c	h_ac	h	h_a	h_c	h_ac	h	h_a
Plug-in	**1.000/0.625**	**1.000/0.625**	**1.000/0.625**	**1.000/0.625**	0.062/4.264	0.062/4.264	0.117/3.802	0.117/3.802	0.224	0.228	0.122	0.117
BFO	0.393/0.377	0.477/0.447	0.749/0.593	0.596/0.577	2.478/2.678	1.609/1.742	**0.801/0.876**	0.936/0.950	109.3	151.8	184.9	146.4
SA	0.874/0.342	1.000/0.315	1.000/0.396	1.000/0.315	0.793/12.09	0.000/12.86	0.004/1.367	0.000/1.099	7.674	0.366	0.515	0.366

Table 3 Training methods comparison for the Breast Cancer dataset

	accuracy train/test				log_loss train/test				fit_time			
h type	h_c	h_ac	h	h_a	h_c	h_ac	h	h_a	h_c	h_ac	h	h_a
Plug-in	1.000/0.373	1.000/0.373	1.000/0.373	1.000/0.373	0.000/22.61	0.000/22.61	0.000/22.61	0.000/22.61	0.917	0.963	0.500	0.550
BFO	1.000/0.373	1.000/0.682	1.000/0.373	1.000/0.622	0.000/22.61	0.933/22.61	0.000/22.61	0.000/22.61	4.732	4.751	4.687	4.932
SA	1.000/0.802	1.000/0.893	**1.000/0.905**	1.000/0.896	18.17/22.61	22.61/22.61	22.61/22.61	22.61/22.61	2.360	2.386	2.347	2.394

Table 4 Training methods comparison for the Wine dataset

	accuracy train/test				log_loss train/test				fit_time			
h type	h_c	h_ac	h	h_a	h_c	h_ac	h	h_a	h_c	h_ac	h	h_a
Plug-in	1.000/0.331	1.000/0.331	1.000/0.331	1.000/0.331	0.000/1.099	0.000/1.099	0.000/1.099	0.000/1.099	0.577	0.607	0.193	0.192
BFO	0.994/0.579	1.000/0.730	1.000/0.331	1.000/0.708	**0.100/1.002**	0.000/1.065	0.000/1.099	0.000/1.075	18.35	0.118	0.122	0.124
SA	1.000/0.725	1.000/0.726	1.000/0.725	**1.000/0.742**	0.339/1.044	**0.262/1.002**	0.189/1.013	0.324/0.974	0.162	0.058	0.060	0.060

Table 5 Training methods comparison for the ILPD dataset

	accuracy train/test				log_loss train/test				fit_time			
h type	h_c	h_ac	h	h_a	h_c	h_ac	h	h_a	h_c	h_ac	h	h_a
Plug-in	1.000/0.715	1.000/0.715	**1.000/0.732**	**1.000/0.732**	0.000/10.27	0.000/10.27	0.000/10.27	0.000/10.27	0.308	0.323	0.159	0.156
BFO	0.999/0.694	0.999/0.670	1.000/0.715	1.000/0.706	1.209/10.18	3.072/10.19	0.000/10.27	1.026/10.27	97.70	92.98	1.838	1.812
SA	1.000/0.600	1.000/0.575	1.000/0.663	1.000/0.639	2.922/10.09	**7.543/9.810**	4.569/9.992	6.812/9.907	0.908	0.895	0.908	0.896

Table 6 Summary of best-performing approaches for PNN training for each benchmarking dataset

Dataset	Number of features	Accuracy rank	Loss rank
Iris	4	Plug-in/BFO	BFO
Ghosts, Goblins, and Ghouls	5	Plug-in	BFO
ILPD	10	Plug-in	SA
Wine	13	SA	*SA/BFO*
Breast Cancer	30	SA	*Inconclusive*

A method was considered as best-performing based on the cross-validation metric value on the test set from Tables 1, 2, 3, 4 and 5. If scores values are the same for all methods the ranking was resolved as *Inconclusive*

optimisation procedure. Especially, the SA algorithm has produced high values of test cross-validation metrics. In the end, a comparison of training methods for the ILPD dataset is presented in Table 5. Here, all algorithms presented approximately similar results, with a slightly higher test accuracy for the Plug-in.

Table 6, shows the best method for various datasets in terms of cross-validation *accuracy* and *log loss* score on the test set. Additionally, the number of features for each dataset was highlighted in the second column of Table 6. It might be noticed, that metaheuristic techniques tend to achieve better performance than classical methods for tasks with a higher number of features, particularly the SA algorithm,. To inspect this trend further experiments should be conducted.

6 Summary and Conclusions

This study conducted a comparison between nature-inspired Simulated Annealing and Bacterial Foraging optimisation techniques and the analytical Plug-in procedure for the traditional training of Probabilistic Neural Networks (PNNs). Gaussian kernels were employed in the PNN pattern layer, and the metaheuristics were specifically tailored for this task, as detailed in Sects. 3 and 4.

Tables 1, 2, 3, 4 and 5 present the comparison results for the mentioned PNN training procedures. In general, BFO and SA exhibited performance similar to the Plug-in method across benchmark datasets, highlighting their potential as optimisation techniques for PNNs. Notably, the SA algorithm outperformed other methods significantly for the Wine and Cancer datasets, suggesting a promising avenue for enhancing PNN applications in similar tasks.

For future research, a more in-depth evaluation of SA and BFO methods could be conducted to identify superior performance on specific datasets. Especially, the relationship between a number of features in data and the SA performance might be explored. Additionally, fine-tuning the parameters of these techniques may enhance their efficacy. Furthermore, exploring the application of these computational intelligence approaches for training advanced versions of PNNs could be a fruitful avenue for investigation.

Acknowledgements This work was partially supported by the program "Excellence initiative—research university" for the AGH University of Krakow and by Grant for Statutory Activity from the Faculty of Physics and Applied Computer Science of the AGH University of Krakow.

References

1. AL-Hadi, I.A.A., Hashim, S.Z.M., Shamsuddin, S.M.H.: Bacterial foraging optimization algorithm for neural network learning enhancement. In: 2011 11th international conference on hybrid intelligent systems (HIS), pp 200–205
2. Ciarelli, P., Krohling, R., Oliveira, E.: Particle swarm optimization applied to parameters learning of probabilistic neural networks for classification of economic activities. In: Particle Swarm Optimization, chap 19, pp. 313–328. InTech (2009)
3. Dua, D., Graff, C.: UCI machine learning repository (2017). http://archive.ics.uci.edu/ml
4. Kan, W.: Ghouls, goblins, and ghosts... boo! (2016). https://kaggle.com/competitions/ghouls-goblins-and-ghosts-boo
5. Kirkpatrick, S., Gelatt, C., Vecchi, M.: Optimization by simulated annealing. Science (New York, N.Y.) 220:671–680 (1983)
6. Kowalski, P.A., Kusy, M., Kubasiak, S., Łukasik, S.: Probabilistic neural network—parameters adjustment in classification task, pp. 1–8 (2020)
7. Kowalski, P.A., Wadas, K.: Triggering probabilistic neural networks with flower pollination algorithm. In: Computational Intelligence and Mathematics for Tackling Complex Problems (2019)
8. Kulczycki, P.: Kernel Estimators in Industrial Applications, vol. 226, pp. 69–91. Springer, Berlin, Heidelberg (2008)
9. Kusy, M., Zajdel, R.: Application of reinforcement learning algorithms for the adaptive computation of the smoothing parameter for probabilistic neural network. IEEE Trans. Neur. Netw. Learn. Syst. **26**(9), 2163–2175 (2015)
10. Naik, S.M., Jagannath, R.P.K., Kuppili, V.: Bat algorithm-based weighted Laplacian probabilistic neural network. Neur. Comput. Appl. **32**, 1157–1171 (2020)
11. Passino, K.: Biomimicry of bacterial foraging for distributed optimization and control. IEEEE Control Syst. Mag. **22**, 52–67 (2002)
12. Pedregosa, F., Varoquaux, G., Gramfort, A., Michel, V., Thirion, B., Grisel, O., Blondel, M., Prettenhofer, P., Weiss, R., Dubourg, V., Vanderplas, J., Passos, A., Cournapeau, D., Brucher, M., Perrot, M., Duchesnay, E.: Scikit-learn: machine learning in Python. J. Mach. Learn. Res. **12**, 2825–2830 (2011)
13. Rere, L., Fanany, M.I., Arymurthy, A.: Simulated annealing algorithm for deep learning, vol. 72 (2015)
14. Shah, H.: Using new artificial bee colony as probabilistic neural network for breast cancer data classification. Front. Eng. Built Environ. ahead-of-print (2021)
15. Specht: Probabilistic neural networks for classification, mapping, or associative memory. In: IEEE 1988 International Conference on Neural Networks, vol. 1, pp. 525–532 (1988)

Inheritance of Completeness Between Systems of Strong and Weak Implications

Francisco Pérez-Gámez, Carlos Bejines, Pablo Cordero, Domingo López-Rodríguez, and Manuel Ojeda-Hernández

Abstract The study of unknown information in formal contexts can be done from two extremely different points of view: working just with the information available at the moment, or exploring all the different values that the unknown information can take. From these two perspectives, we obtain two kinds of attribute implications: the weak ones which are the attribute implications that hold with the current amount of information, and the strong ones which will also hold under any update of the context. We study whether, given a complete system of weak implications concerning partial formal context, one can extract a complete system of strong ones concerning the same partial formal context.

Keywords Formal concept analysis · Unknown information · Attribute implication · Completeness

1 Introduction

The exploration of unknown information in Formal Concept Analysis (FCA) holds significant importance, as evidenced in [1–3]. This exploration involves two distinct perspectives: one that considers only the available information and another that delves into all possible scenarios for missing information. In [4], the authors explored the former approach, that is, they studied the implications that hold with the known information and proved that these implications may lose validity when confronted with new information. They thus coined the term weak implications.

Similarly, the exploration of logical rules that remain true in the presence of new data is also of importance. These are the so-called strong implications. Obtaining this kind of implication requires an exhaustive examination of all conceivable scenarios.

From a practical standpoint, this exploration leads to the enumeration of all possible scenarios, whose amount increases exponentially with the size of the context and the amount of unknown information. Hence, the computational burden increases

F. Pérez-Gámez · C. Bejines · P. Cordero · D. López-Rodríguez (✉) · M. Ojeda-Hernández
University of Malaga, Malaga, Spain
e-mail: dominlopez@uma.es

L. T. Kóczy et al. (eds.), *Computational Intelligence and Mathematics for Tackling Complex Problems 6*, Studies in Computational Intelligence 1222,
https://doi.org/10.1007/978-3-031-97879-1_20

with the volume of data. Consequently, while strong implications offer more knowledge, practical considerations often lead to a preference for working with weak implications due to their relative ease of derivation.

In the broader context of working with attribute implications, it becomes crucial to establish a sound subset of implications, that captures all the knowledge within the context. A well-known example of this is the Duquenne-Guigues basis [5] which is widely recognized in FCA and serves as a suitable choice for this purpose.

In this paper, we compare the two kinds of attribute implications and focus on the question of whether the completeness of a system of weak implications concerning a partial formal context is inherited from a subset of strong ones. The structure of the paper is the following: Sect. 2 provides preliminary results, Sect. 3 explores the relation between both kinds of implications, and Sect. 4 presents the conclusions and future works.

2 Preliminaries

We assume that the reader is familiar with the basic notions of lattice theory [6, 7] and formal concept analysis [8, 9], and we summarize below only the concepts and results we are going to use on formal concept analysis with unknown information developed in the work [4].

2.1 The $\dot{\mathbf{3}}^M$ Sets

Classical sets can be viewed as functions from the universe to the two-values set $\{0, 1\}$ using the so-called characteristic mapping. We extend this idea by considering the set of four values $\mathbf{4} = \{\circ, +, -, \iota\}$ where $\circ$ denotes missing information, $+$ denotes positive information, $-$ denotes negative information, and ι denotes contradictory information. In addition, we consider the order $\leq$ over $\mathbf{4}$ given by the lattice in Fig. 1.

Thus, given a non-empty universe M, we denote by $\mathbf{4}^M$ the set of $\mathbf{4}$-valued sets, i.e. the mappings $A : M \to \mathbf{4}$. In addition, the pointwise extension of $(\mathbf{4}, \leq)$ to $\mathbf{4}^M$ is a complete lattice. The pointwise extension of $\leq$ will be denoted by $\sqsubseteq$, that is,

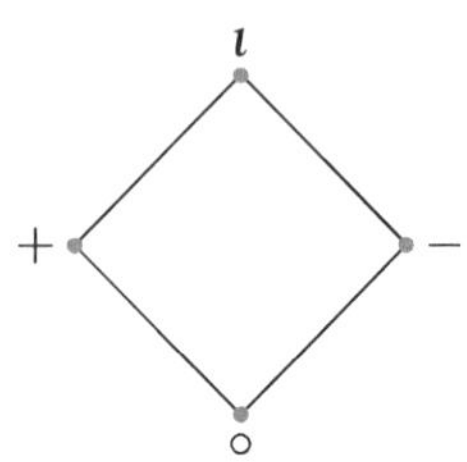

Fig. 1 The lattice $(\mathbf{4}, \leq)$

$A \sqsubseteq B$ iff $A(m) \leq B(m)$ for all $m \in M$. Its bottom element is denoted by ε having that $\varepsilon(m) = \circ$ for all $m \in M$ whereas its top element is denoted by $\dot{\iota}$ having that $\dot{\iota}(m) = \iota$ for all $m \in M$.

As in the classical case, in the propositional logic that we work with, from contradictory information, we can derive everything. As a consequence, we define the following closure operator

$$\mathcal{O}\colon \mathbf{4}^U \to \mathbf{4}^U \text{ being } \mathcal{O}(A) = \begin{cases} A & \text{if } A(m) \neq \iota \text{ for all } m \in M, \\ \dot{\iota} & \text{otherwise.} \end{cases}$$

The codomain of $\mathcal{O}$, denoted by $\dot{\mathbf{3}}^M$, is a closure system and, therefore, a $\wedge$-subsemilattice of $(\mathbf{4}^M, \sqsubseteq)$ and a complete lattice. The join and meet operations of the complete lattice $(\dot{\mathbf{3}}^M, \sqsubseteq)$ will be denoted by $\sqcup$ and $\sqcap$ respectively.

2.2 *Partial Formal Contexts and Attribute Implications*

A **partial formal context** is a triple $\mathbb{P} = (G, M, I)$ where G and M are not-empty sets, whose elements are called objects and attributes, respectively, and I is a three-valued relation $G \times M \to \mathbf{4} \setminus \{\iota\}$. Given $g \in G$ and $m \in M$, $I(g, m) = +$ means that *"we have information that the object g has the attribute m"*, $I(g, m) = -$ means that *"we have information that the object g does not have the attribute m"* and $I(g, m) = \circ$ means that *"we have no information about whether the object g has the attribute m or not"*.

As mentioned in the introduction, the study of unknown information is two-fold. Each partial formal context induces two Galois connections and two different definitions of attribute implication. Given space limitations, we present only the notions we need to present these kinds of implications. First, We present the following operators $(\ )^{\uparrow} : \mathbf{2}^G \to \dot{\mathbf{3}}^M$ and $(\ )^{\downarrow} : \dot{\mathbf{3}}^M \to \mathbf{2}^G$ defined as

$$X^{\uparrow} = \bigsqcap_{g \in X} I(g, \ ), \quad \text{and} \quad Y^{\downarrow} = \{g \in G \mid g^{\uparrow} \sqcap Y = Y\}$$

These operators form a Galois connection.

Definition 1 ([4]) Given a partial formal context $\mathbb{P} = (G, M, I)$ and $A, B \in \dot{\mathbf{3}}^M$, we say that $\mathbb{P}$ is a **model of the weak implication** $A \rightsquigarrow B$, denoted by $\mathbb{P} \models A \rightsquigarrow B$, if

$$B \sqsubseteq \bigsqcap_{\{g \in G \mid A \sqsubseteq g^{\uparrow}\}} g^{\uparrow} = A^{\downarrow\uparrow}.$$

In addition, a partial formal context $\mathbb{P}$ is said to be a model for a set of weak implications Σ, denoted by $\mathbb{P} \models \Sigma$, if $\mathbb{P} \models A \rightsquigarrow B$ for all $A \rightsquigarrow B \in \Sigma$.

As usual, this semantic definition leads to the notion of semantic entailment. A weak implication $A \rightsquigarrow B$ is semantically entailed from a set of weak implications Σ when for all partial formal context $\mathbb{P}$, we have that $\mathbb{P} \models \Sigma$ implies $\mathbb{P} \models A \rightsquigarrow B$. It is denoted by $\Sigma \models A \rightsquigarrow B$.

In [4], we provide the following Armstrong-style axiomatic system for weak implications: for all $A, B, C \in \dot{\underline{3}}^M$,

[Inc] **Inclusion**: Infer $A \sqcup B \rightsquigarrow A$.
[Augm] **Augmentation**: From $A \rightsquigarrow B$ infer $A \sqcup C \rightsquigarrow B \sqcup C$.
[Trans] **Transitivity**: From $A \rightsquigarrow B$ and $B \rightsquigarrow C$ infer $A \rightsquigarrow C$.

Thus, a weak implication $A \rightsquigarrow B$ is said to be syntactically derived from a set of weak implications Σ, denoted by $\Sigma \vdash_w A \rightsquigarrow B$, if there is a sequence of weak implications $A_1 \rightsquigarrow B_1, ..., A_n \rightsquigarrow B_n$ such that $A_n \rightsquigarrow B_n = A \rightsquigarrow B$ and, for all $1 \leq i \leq n$, one of the following conditions is satisfied: $A_i \rightsquigarrow B_i \in \Sigma$, $A_i \rightsquigarrow B_i$ is an axiom, or $A_i \rightsquigarrow B_i$ is obtained from implications belonging to $\{A_j \rightsquigarrow B_j \mid 1 \leq j < i\}$ by applying the above inferences rules. In [4], the soundness and completeness of this axiomatic system (i.e. $\Sigma \models A \rightsquigarrow B$ iff $\Sigma \vdash_w A \rightsquigarrow B$) was proved.

On the other hand, there are implications that hold in all the possible scenarios for the unknown information, these are the so-called strong implications. In order to collect all the possible information, in [10] the authors present and order between the partial formal contexts denoted by $\preceq$. With this order, given a partial formal context, we present the following set of partial formal contexts which form a complete lattice with the order $\preceq$: $\mathfrak{P}(\mathbb{P}_0) = \{\mathbb{P} : \mathbb{P} \preceq \mathbb{P}_0\}$.

With the complete lattice and with $\dot{\mathbf{3}}^M$ we can define the following derivation operators to collect all the possible information from a partial formal context:

- $^{\Uparrow} : \mathfrak{P}(\mathbb{P}_0) \to \dot{\mathbf{3}}^M$ that maps any $\mathbb{X} = (G, M, I) \in \mathfrak{P}(\mathbb{P}_0)$ to

$$\mathbb{X}^{\Uparrow} = \bigwedge_{g \in G} I(g, \ ).$$

- $^{\Downarrow} : \dot{\mathbf{3}}^M \to \mathfrak{P}(\mathbb{P}_0)$ that maps any $\dot{\mathbf{3}}$-set $A \in \dot{\mathbf{3}}^M$ to $A^{\Downarrow} = (G, M, I)$ where

$$G = \{g \in G_0 \colon I_0(g, \ ) \sqcup A \neq i\} \text{ and}$$
$$I(g, \ ) = I_0(g, \ ) \sqcup A, \text{ for each } g \in G.$$

First, we present the operators that allow us to capture all the possible information in the partial formal context:

Definition 2 ([11]) Given a partial formal context $\mathbb{P}$ and $A, B \in \dot{\mathbf{3}}^M$, we say that $\mathbb{P}$ is a **model of the strong implication** $A \Rightarrow B$, denoted by $\mathbb{P} \models A \Rightarrow B$ if

$$B \sqsubseteq \bigsqcap_{\{g \in G \mid g^{\uparrow} \sqcup A \neq i\}} (g^{\uparrow} \sqcup A) = A^{\Downarrow\Uparrow}.$$

For this new type of implication, the $\vdash_s$ inference system is still correct but not complete. As it was proved in [11], to achieve completeness it becomes necessary to add the following two inference rules: for all $A \in \underline{\dot{\mathbf{3}}}^M$ and all singletons $a, b \in \underline{\dot{\mathbf{3}}}^M$,

[Rft] **Reflection**: From $A \sqcup a \Rightarrow b$ infer $A \sqcup \overline{b} \Rightarrow \overline{a}$.
[Tru] **Trust**: From $a \Rightarrow \overline{a}$ infer $\varepsilon \Rightarrow \overline{a}$.

where a singleton is a set $x \in \underline{\dot{\mathbf{3}}}^M$ such that $x(m) = \circ$ for all $m \in M$ except for an m_0 and, furthermore, $x(m_0) = +$ or $x(m_0) = -$. Moreover, $\overline{x}$ is the singleton obtained from x by exchanging $+$ for $-$ and vice versa.

The syntactic derivation induced by the new set of inference rules will be denoted by $\vdash_s$.

Finally, given a partial formal context, we define the notion of a sound and complete set of implications concerning that partial formal context:

Definition 3 Let $\mathbb{P}$ be a partial formal context.

A set of weak implications Σ is said to be sound and complete concerning $\mathbb{P}$ when $\mathbb{P} \models A \rightsquigarrow B$ if and only if $\Sigma \vdash_w A \rightsquigarrow B$ for all weak implication $A \rightsquigarrow B$.

A set of strong implications Σ is said to be sound and complete concerning $\mathbb{P}$ when $\mathbb{P} \models A \Rightarrow B$ if and only if $\Sigma \vdash_s A \Rightarrow B$ for all strong implication $A \Rightarrow B$.

3 Relationship Between Both Kinds of Implications

Computing a sound and complete set of weak implications for a given partial formal context is less computationally expensive than doing so for strong implications. In this section, we study whether it is possible to obtain the latter from the former. The proofs have been omitted due to space constraints and will be presented in the extended version.

Theorem 1 *Let $\mathbb{P}$ be a partial formal context and $A, B \in \dot{\mathbf{3}}^M$.*

1. *$\mathbb{P} \models A \Rightarrow B$ implies that $\mathbb{P} \models A \rightsquigarrow B$.*
2. *If $\mathbb{P} \models A \rightsquigarrow B$ but $\mathbb{P} \not\models A \Rightarrow B$ then there exist at least one object $g \in G$ such that $I(g, m) = \circ$, $A(m) \neq \circ$ and $B \not\sqsubseteq g^{\uparrow} \sqcup A$.*

Observe that the converse of the first assertion does not hold. Indeed, the second assertion gives a necessary condition for a valid weak implication not to be a strong one. In addition, the following example shows a weak implication that is not strong for a given partial formal context. The lack of strength is shown by using the characterization in the last result.

Example 1 Consider the partial formal context $\mathbb{P}$ given by the next table:

$\mathbb{P}$	a	b	c
o_1	$\circ$	$-$	$-$
o_2	$\circ$	$+$	$+$
o_3	$-$	$+$	$\circ$

It is a matter of calculation to prove that $\mathbb{P} \models \overline{a} \rightsquigarrow b$ but $\mathbb{P} \not\models \overline{a} \Rightarrow b$. Here, $I(o_i, a) = \circ$ for $i = 1, 2$. We have that $\mathbb{P} \not\models \overline{a} \Rightarrow b$ because $b \not\sqsubseteq (o_1^\uparrow \sqcup a) \sqcap (o_2^\uparrow \sqcup a) = a^{\Downarrow\Uparrow}$. Observe that o_2 does not affect to the validity of the strong implication since $b \sqsubseteq (o_2^\uparrow \sqcup a)$.

Our problem now is, given a partial formal context $\mathbb{P}$, to check if having a sound and complete set of weak implications concerning $\mathbb{P}$ we can extract a sound and complete set of strong implications concerning $\mathbb{P}$. First, we need to look for a way to, given a partial formal context, build a sound and complete set of weak implications concerning that partial formal context. To achieve this goal, we consider some definitions that extend the known notions of quasi and pseudo-intent [12] to the framework of weak implications.

Definition 4 Let $B \in \dot{\mathbf{3}}^M$, we say that it is a weak quasi-intent if for all $X \sqsubseteq B$ we have that $X^{\downarrow\uparrow} \sqsubset B$ or $X^{\downarrow\uparrow} = B^{\downarrow\uparrow}$. We denote the set of all weak quasi-intents by $\mathbb{Q}_W$.

Definition 5 We say that $B \in \mathbb{Q}_W$ is a weak pseudo-intent if $B^{\downarrow\uparrow} \neq B$ and it holds that for all $X \in \mathbb{Q}_W, X \sqsubset B$ we have that $X^{\downarrow\uparrow} \sqsubset B$. The set of all weak pseudo-intents is denoted by $\mathcal{P}_W$.

Finally, we can build the desired set of weak implications.

Proposition 1 *Let $\mathbb{P}$ be a partial formal context, the following set of weak implications is sound and complete concerning $\mathbb{P}$:*

$$\left\{A \rightsquigarrow A^{\downarrow\uparrow} \mid A \in \mathcal{P}_W\right\}.$$

The proof is skipped since it is analogous to the one of Theorem 7, in [8, Chap. 3] using the results on the unknown-information framework of FCA introduced in [4]. The answer to the research question posed in this paper is negative. The following example shows a counterexample of the claim.

Example 2 Let $\mathbb{P} = (G, M, I)$ be the partial formal context given by the next table:

$\mathbb{P}$	a	b
g_1	$\circ$	$+$
g_2	$-$	$\circ$

By using Proposition 1, we build the following set of weak implications which is sound and complete concerning $\mathbb{P}$:

$$\{a \rightsquigarrow i, \overline{b} \rightsquigarrow i, \overline{a}b \rightsquigarrow i\}.$$

None of these weak implications is a valid strong implication. However, as we can see $b \sqsubseteq a^{\Downarrow\Uparrow} = g_1^\uparrow \sqcup a = ab$ and, by definition, we have that $a \Rightarrow b$ is a valid implication.

Thus, given a partial formal context $\mathbb{P}$, it is not always possible to obtain a sound and complete set of strong implications concerning $\mathbb{P}$ from a sound and complete set of weak implications concerning $\mathbb{P}$.

4 Conclusion and Further Works

In this paper, we have studied the properties of the strong implications in the framework of partial formal contexts. The knowledge we get from these implications holds whenever the information in the context is updated. Specifically, we have proved that every strong implication is weak and the converse does not hold. Furthermore, we have provided a characterization for discerning whether a valid weak implication is also strong. From the application point of view, it would greatly simplify the cost of computation to get a complete set of strong implications concerning a partial formal context from a weak one. Unfortunately, the example provided shows that this is not possible in general. As a future work, we will study the properties that need to be satisfied by the context to make this possible and also search for new less-costly strategies to compute a complete set of strong implications.

Acknowledgements This work has been partially funded by the State Agency of Research (AEI), the Ministerio de Ciencia, Innovación y Universidades (MCIU), the European Social Research Fund (FEDER), the Junta de Andalucía (JA), and the Universidad de Málaga (UMA) through the PhD contract FPU19/01467 (MCIU), the VALID research project (PID2022-140630NB-I00 funded by MCIN/AEI/10.13039/501100011033) and the research project PID2021-127870OB-I00 (MCIU/AEI/FEDER, UE).

References

1. Lang, G., Yao, Y.: Formal concept analysis perspectives on three-way conflict analysis. Int. J. Approx. Reason **152**, 160–182 (2023)
2. Felde, M., Stumme, G.: Interactive collaborative exploration using incomplete contexts. Data Knowl. Eng. **143**, 102104 (2023)
3. Dubois, D., Medina, J., Prade, H., Ramírez-Poussa, E.: Disjunctive attribute dependencies in formal concept analysis under the epistemic view of formal contexts. Inf. Sci. **561**, 31–51 (2021)
4. Pérez-Gámez, F., Cordero, P., Enciso, M., Mora, Á.: Simplification logic for the management of unknown information. Inf. Sci. **634**, 505–519 (2023)
5. Guigues, J., Duquenne, V.: Familles minimales d'implications informatives résultant d'une tables de données binaires. Mathématiques et Sciences Humaines **95**, 5–18 (1986)
6. Birkhoff, G.: Lattice Theory, 1st edn. American Mathematical Society, Providence (1940)
7. Davey, B., Priestley, H.: Introduction to Lattices and Order, 2nd edn. Cambridge University Press, Cambridge (2002)
8. Ganter, B., Obiedkov, S.: Conceptual Exploration. Springer, Berlin (2016)
9. Ganter, B., Wille, R.: Formal Concept Analysis. Mathematical Foundations. Springer, Berlin (1996)
10. Perez-Gamez, F., Cordero, P., Enciso, M., Mora, A.: A Galois connection between partial formal contexts and attribute sets. In: The 16th International Conference on Concept Lattices and Their Applications, pp. 45–55 (2022)
11. Pérez-Gámez, F., Cordero, P., Enciso, M., Mora, Á.: Defining strong implications in partial formal context, Submitted to Journal
12. Kuznetsov, S.O., Obiedkov, S.A.: Some decision and counting problems of the Duquenne-Guigues basis of implications. Discr. Appl. Math. **156**(11), 1994–2003 (2008)

Migrative Properties for Triangular Conorms and Fuzzy Implications

M. Eugenia Cornejo, Jesús Medina, and Francisco José Ocaña

Abstract Migrative and cross-migrative properties of t-norms have widely been analyzed in the literature. However, few works exist on migrative properties on t-conorms over fuzzy implications. This paper presents different features and remarks of these properties, which also justify the interconnected relationship between them.

Keywords Migrativity · Cross-migrativity · t-conorm · Fuzzy implication

1 Introduction

Several families of t-norms related to functional equations have been proposed to study certain characteristics and relationships between these operators [1, 10–12, 18, 19]. In this context, Durante and Sarkoci [6] proposed a novel property on binary functions and based on a functional equation, called α-migrativity, to study the construction of t-norms by means of convex combinations of two t-norms.

Migrative properties between other operators, such as aggregations [4, 5] containing t-conorms [9, 14], uninorms [20], nullnorms [22], overlap functions [16] and fuzzy implications [3] have been a widely studied research topic. For example, motivated by the natural duality between t-norms and t-conorms, Baczynski et al. [3] formulated a new description about α-migrativity of t-conorms. Moreover, taking into account that the (S, N)-implications [2] are derived from t-conorms and fuzzy negations, Pan et al. [15] incorporated fuzzy implications into the migrativity equation of t-norms, providing a new definition of the migrativity for t-conorms over fuzzy implications.

M. E. Cornejo (✉) · J. Medina · F. J. Ocaña
Department of Mathematics, University of Cádiz, Cádiz, Spain
e-mail: mariaeugenia.cornejo@uca.es

J. Medina
e-mail: jesus.medina@uca.es

F. J. Ocaña
e-mail: franciscojose.ocana@uca.es

L. T. Kóczy et al. (eds.), *Computational Intelligence and Mathematics for Tackling Complex Problems 6*, Studies in Computational Intelligence 1222,
https://doi.org/10.1007/978-3-031-97879-1_21

On the other hand, following with the study of different classes of t-norms which are closely related to the functional equations, Fodor et al. [8] proposed a weaker form of the classical commuting equation [17], which led to a new property between t-norms, called α-cross-migrativity. This new functional equation was presented as a further generalization of the migrativity equation of t-norms. Moreover, taking into account the achievements previously reached in this context, Fang [7] recently studied the α-cross-migrativity involving t-conorms and fuzzy implications, and proposed a functional equation to characterize the α-cross-migrativity of t-conorms over fuzzy implications, which was studied in more detail in [21].

Besides the previous works, to the best of our knowledge, the literature about the migrative and cross migrative properties between disjunctions and fuzzy implications is really limited, these properties are usually studied independently of each other and there are still open questions to be solved. Taking into account these reasons, the purpose of the present paper is to go further in the study of these properties introducing new motivations, features and examining the relationship among the migrativity and cross-migrativity of t-conorms over fuzzy implications.

2 Preliminaries

This section deals with the basic notions of t-conorms, fuzzy negations and fuzzy implications, which are logical connectives extensively investigated in the literature, in order to make the paper self-contained [2, 13].

Definition 1 Let $S, I\colon [0,1]^2 \to [0,1]$ be two binary mappings and $N\colon [0,1] \to [0,1]$ be a unary mapping. We say that:

- S is a *triangular conorm (t-conorm)* if it is commutative, associative, increasing in both arguments and has 0 as neutral element.
- N is a *fuzzy negation* if it is decreasing and the equalities $N(0) = 1$ and $N(1) = 0$ hold.
- I is a *fuzzy implication* if it is decreasing in the first argument, increasing in the second argument and the equalities $I(0,0) = 1$, $I(1,1) = 1$ and $I(1,0) = 0$ hold.
- I is a *(S, N)-implication* if there exist a t-conorm S and a fuzzy negation N such that $I(x, y) = S(N(x), y)$, for all $x, y \in [0,1]$.

The following result shows how a fuzzy negation can be defined from a given fuzzy implication.

Lemma 1 *Let $I\colon [0,1]^2 \to [0,1]$ a binary mapping. If I is a fuzzy implication, then the mapping $N_I\colon [0,1] \to [0,1]$ defined as $N_I(x) = I(x,0)$, for all $x \in [0,1]$, is a fuzzy negation.*

Well-known examples of t-conorms and fuzzy implications together with their analytic expressions, for all $x, y \in [0,1]$, are given in Table 1.

Table 1 Examples of t-conorms and fuzzy implications

Name	T-conorm	Fuzzy implication
Łukasiewicz	$S_{LK}(x, y) = \min\{1, x + y\}$	$I_{LK}(x, y) = \min\{1, 1 - x + y\}$
Gödel	$S_{GD}(x, y) = \max\{x, y\}$	$I_{GD}(x, y) = \begin{cases} 1 & \text{if } x \leq y \\ y & \text{otherwise} \end{cases}$
Goguen	$S_{GG}(x, y) = x + y - x \cdot y$	$I_{GG}(x, y) = \begin{cases} 1 & \text{if } x \leq y \\ \frac{y}{x} & \text{otherwise} \end{cases}$
Reichenbach		$I_{RC}(x, y) = 1 - x + x \cdot y$
Kleene-Dienes		$I_{KD}(x, y) = \max\{1 - x, y\}$
Weber		$I_{WB}(x, y) = \begin{cases} y & \text{if } x = 1 \\ 1 & \text{otherwise} \end{cases}$

Applying Lemma 1, we can obtain the fuzzy negations associated with the fuzzy implications collected in Table 1. It is easy to see that these fuzzy negations are defined, for each $x \in [0, 1]$, as follows:

$$N_{LK}(x) = N_{RC}(x) = N_{KD}(x) = 1 - x$$

$$N_{GD}(x) = N_{GG}(x) = \begin{cases} 1 & \text{if } x = 0 \\ 0 & \text{otherwise} \end{cases}$$

$$N_{WB}(x) = \begin{cases} 0 & \text{if } x = 1 \\ 1 & \text{otherwise} \end{cases}$$

These negations are called standard negation, intuitionistic negation and dual intuitionistic negation, respectively. Denoting the standard negation as N_S, we can check that I_{RC} is a (S_{GG}, N_S)-implication and I_{KD} is a (S_{GD}, N_S)-implication.

3 Migrative and Cross-Migrative Properties for t-Conorms over Fuzzy Implications

This section focuses on migrative and cross-migrative properties for t-conorms over fuzzy implications, the former is proposed and analyzed in [3, 15] and the latter is introduced and characterized in [7, 21]. The study of these properties in the literature has been studied independently, without considering the existence of possible connections between them. For this reason, our goal focuses on the study of the relationship between the migrativity and cross-migrativity of t-conorms over fuzzy implications.

Definition 2 Let S be a t-conorm and $\alpha \in [0, 1]$. We say that S is α-migrative, if the equality $S(1-\alpha+\alpha x, y) = S(x, 1-\alpha+\alpha y)$ holds, for all $x, y \in [0, 1]$.

It is convenient to observe that, the α-migrative property for t-conorms is closely related to the Reichenbach fuzzy implication. Indeed, taking into account the formula of the aforementioned fuzzy implication, the equality given in Definition 2 can be rewritten as follows:

$$S(I_{RC}(\alpha, x), y) = S(x, I_{RC}(\alpha, y))$$

This fact motivated the study of the α-migrative property for t-conorms over some specific well-known fuzzy implications such that the Reichenbach, Łukasiewicz and Kleene-Dienes implications, among others. After that, a characterization of the α-migrativity for t-conorms over general fuzzy implications was proposed, which is based in the following definition [15].

Definition 3 Let S be a t-conorm, I be a fuzzy implication.

- Given $\alpha \in [0, 1]$, S is *α-migrative over I*, if $S(I(\alpha, x), y) = S(x, I(\alpha, y))$ holds, for all $x, y \in [0, 1]$.
- S is *migrative over I*, if S is α-migrative over I, for all $\alpha \in [0, 1]$.

The next result associates the migrative property for t-conorms with the class of (S, N)-implications.

Proposition 1 *([15]) Let S be a t-conorm and I be a fuzzy implication. S is migrative over I if and only if I is a (S, N_I)-implication.*

The cross-migrative property for t-conorms over fuzzy implications arises as a novel proposal to find relationships between disjunctions and fuzzy implications [7]. The formal definition is presented below.

Definition 4 Let S be a t-conorm, I be a fuzzy implication.

- Given $\alpha \in [0, 1]$, S is *α-cross-migrative over I*, if $I(x, S(\alpha, y)) = S(y, I(x, \alpha))$ holds, for all $x, y \in [0, 1]$.
- S is *cross-migrative over I*, if S is α-cross-migrative over I, for all $\alpha \in [0, 1]$.

$$I(x, S(\alpha, y)) = S(y, I(x, \alpha))$$

Following the philosophy of Proposition 1, we are also interested in relating the class of (S, N)-implications to the α-cross-migrative property for t-conorms.

Proposition 2 *Let S be a t-conorm and I be a fuzzy implication. S is 0-cross-migrative over I if and only if I is a (S, N_I)-implication.*

Table 2 Formulae associated with the α-cross migrativity and x-migrativity

Property	Fuzzy expression	Boolean formula
α-cross-migrativity	$I(x, S(\alpha, y)) = S(y, I(x, \alpha))$	$x \to (\alpha \vee y) \equiv y \vee (x \to \alpha)$
x-migrativity	$S(I(x, \alpha), y) = S(\alpha, I(x, y))$	$(x \to \alpha) \vee y \equiv \alpha \vee (x \to y)$

Remark 1 Notice that, the definitions of the migrativity and cross-migrativity properties was widely argued in previous papers [3, 7, 15]. They arose from the corresponding properties of migrativity and cross-migrativity of t-norms. Moreover, we can remark here that these properties are also extension of Boolean formulae. Specifically, we can consider a correspondence between the migrative and cross-migrative properties for t-conorms over fuzzy implications and the logic formulae shown in Table 2, interpreting the t-conorm as the disjunction $\vee$ and the fuzzy implication as the classical implication $\to$.

It is easy to check that for each of the above mentioned property, the truth tables of the corresponding logical functions are the same, and hence, the preceding formulae are equivalent in the Boolean case. Therefore, we have also shown that migrative and cross-migrative properties are a generalization of Boolean equivalences. Moreover, we can see in Table 2 that both equivalences are related since, by the commutativity of the disjunction, the expression $y \vee (x \to \alpha)$ in the first equivalence is equal to $(x \to \alpha) \vee y$ of the second equivalence. ■

Next, the main result of the paper is introduced, which shows that the migrativity and cross-migrativity properties for a t-conorm and a fuzzy implication are related as in the Boolean setting.

Theorem 1 *Let S be a t-conorm and I be a fuzzy implication. S is cross-migrative over I if and only if S is migrative over I.*

Thus, a new relationship between migrative and cross-migrative properties have been obtained, which naturally also extend the properties of the Boolean case.

4 Conclusions and Further Work

This paper has presented properties, motivations and relations between the migrative and cross-migrative properties for t-conorms over fuzzy implications. We have emphasized in Remark 1 that the migrative and cross-migrative properties are natural extension from both points of view, the corresponding properties for t-norms and their associated Boolean formulae. However, other extensions and properties are being introduced in the literature without an apparently justification and motivation on a natural property nor based on other classical properties. For example, a new definition of migrativity for aggregation functions was introduced in [5], as a generalization of the α-migrative property for t-norms previously presented in the literature [6, 9]. However, as it is mentioned in Corollary 2 of [5], there are no t-conorms, uninorms or nullnorms that verify the proposed migrativity equation. Thus, the study of properties of general fuzzy operators is important, but its relevance must be clearly motivated and justified. In the future, we will continue inspecting new general properties of fuzzy operators based on well-known classical properties and/or appropriately motivated and justified.

Acknowledgements Partially supported by the project PID2019-108991GB-I00 funded by MICIU/AEI/10.13039/501100011033, the project PID2022-137620NB-I00 funded by MICIU/AEI/10.13039/501100011033 and FEDER, UE, by the grant TED2021-129748B-I00 funded by MCIN/AEI/10.13039/501100011033 and European Union NextGenerationEU/PRTR, and by the industrial predoctoral contract PU/EPIF-FPI-GRUPOENERGETICOPUERTOREAL/CP/2022-051, corresponding to the Research and Transfer Promotion Program of the University of Cádiz 2018/2019.

References

1. Aczél, J., Alsina, C.: Characterizations of some classes of quasilinear functions with applications to triangular norms and to synthesizing judgements. Aequationes mathematicae **25**, 313–315 (1982)
2. Baczyński, M., Jayaram, B.: Fuzzy Implications, vol. 231. Springer (2008)
3. Baczyński, M., Jayaram, B., Mesiar, R.: Fuzzy implications: alpha migrativity and generalised laws of importation. Inf. Sci. **531**, 87–96 (2020)
4. Bustince, H., De Baets, B., Fernandez, J., Mesiar, R., Montero, J.: A generalization of the migrativity property of aggregation functions. Inf. Sci. **191**, 76–85 (2012)
5. Bustince, H., Montero, J., Mesiar, R.: Migrativity of aggregation functions. Fuzzy Sets Syst. **160**, 766–777 (2009)
6. Durante, F., Sarkoci, P.: A note on the convex combinations of triangular norms. Fuzzy Sets Syst. **159**, 77–80 (2008)
7. Fang, B.W.: On alpha-cross-migrativity of t-conorms over fuzzy implications. Fuzzy Sets Syst. **466** (2023)
8. Fodor, J., Klement, E.P., Mesiar, R.: Cross-migrative triangular norms. Int. J. Intell. Syst. **27**, 411–428 (2012)
9. Fodor, J., Rudas, I.J.: On continuous triangular norms that are migrative. Fuzzy Sets Syst. **158**, 1692–1697 (2007)
10. Fodor, J.C., Keresztfalvi, T.: A characterization of the Hamacher family of t-norms. Fuzzy Sets Syst. **65**(1), 51–58 (1994)
11. Frank, M.J.: On the simultaneous associativity ot $f(x, y)$ and $x + y - f(x, y)$. Aequationes Mathematicae **19**, 194–226 (1979)
12. Klement, E.P., Mesiar, R., Pap, E.: A characterization of the ordering of continuous t-norms. Fuzzy Sets Syst. **86**(2), 189–195 (1997)
13. Klement, E.P., Mesiar, R., Pap, E.: Triangular norms. Position paper I: Basic analytical and algebraic properties, vol. 143, pp. 5–26. Elsevier B.V. (2004)
14. Mas, M., Monserrat, M., Ruiz-Aguilera, D., Torrens, J.: On migrative t-conorms and uninorms. In: International Conference on Information Processing and Management of Uncertainty (2012)
15. Pan, D., Zhou, H., Yan, X.: Characterizations for the migrativity of continuous t-conorms over fuzzy implications. Fuzzy Sets Syst. **456**, 173–196 (2023)
16. Qiao, J., Hu, B.Q.: On generalized migrativity property for overlap functions. Fuzzy Sets Syst. **357**, 91–116 (2019)
17. Saminger-Platz, S., Mesiar, R., Dubois, D.: Aggregation operators and commuting. IEEE Trans. Fuzzy Syst. **15**, 1032–1045 (2007)

18. Saminger-Platz, S., Sarkoci, P., Baets, B.D.: The dominance relation on the class of continuous t-norms from an ordinal sum point of view (2006)
19. Schweizer, B., Sklar, A.: Probabilistic Metric Spaces. Courier Corporation (2011)
20. Su, Y., Zong, W., Wen Liu, H., Zhang, F.: On migrativity property for uninorms. Inf. Sci. **300**, 114–123 (2015)
21. Wang, C.Y., Wan, L., Zhang, B.: Notes on alpha-cross-migrativity of t-conorms over fuzzy implications. Fuzzy Sets Syst. **473**, 108741 (2023)
22. Zong,W., Su, Y., Liu, H.: Migrative property for nullnorms. Int. J. Uncertain. Fuzziness Knowl. Based Syst. **22**, 749–760 (2014)

Masonry Strength Assessment Based on Fuzzy Signature Model

A. Kaszás, A. Dormány, V. Pomezanski, Z. Orbán, and L. T. Kóczy

Abstract Fuzzy signatures have been successfully used for various engineering applications including the strength and condition assessment of structural materials. In this paper, fuzzy signatures were used to determine the compressive strength of masonry based on groups of related measured values. Handling the uncertainty this way seemed useful because of the subjective parts and influencing noise factors of the measurements which are incorporated as leaves in the signature. For such modeling the structure from where the algebraic framework can be obtained which allows making computations with the fuzzy signatures thus determined. Since multiplicative type aggregations are applied on the various material test results assigned to the leaves of the signatures for the determination of the compressive strength of masonry, fuzzy arithmetic multiplication based on Zadeh's extension principle was applied here. To perform the fuzzy signature calculations, the scalar product and the n-th root of fuzzy numbers were used.

Keywords Fuzzy signatures · Weighted geometric aggregations · Compressive strength · Masonry

1 Introduction

In various scientific areas, handling non-probabilistic uncertainties is not avoidable. In many practical problems, the uncertainty comes from linguistic expressions representing assessment of the systems under investigation, which cannot be adequately

A. Kaszás (✉)
Faculty of Engineering and Information Technologies, University of Pécs, Pécs, Hungary
e-mail: kaszas.andras@mik.pte.hu

A. Dormány · V. Pomezanski · Z. Orbán
Structural Diagnostics and Analyses Research Group, Faculty of Engineering and Information Technologies, University of Pécs, Pécs, Hungary

L. T. Kóczy
Dept. of Information Technology, Széchenyi István University Győr, Győr, Hungary

L. T. Kóczy et al. (eds.), *Computational Intelligence and Mathematics for Tackling Complex Problems 6*, Studies in Computational Intelligence 1222,
https://doi.org/10.1007/978-3-031-97879-1_22

modeled with statistical expressions. These expressions are based on (often subjective) human judgments represented within technical terms describing the given problems, but they are inaccurate in the mathematical meaning. Despite these expressions describing a problem just qualitatively, it is possible to assign values to them and to create a solvable mathematical problem (see e.g. [1, 2]).

The determination of the load-bearing capacity of structures requires knowledge of material properties in addition to geometric characteristics. In the case of masonry structures, compressive strength is considered the most important material property. Assessing the compressive strength of an existing masonry structure is a challenging engineering task, because many times the reliable destructive testing methods cannot be applied. Therefore, less reliable non-destructive testing methods need to be used. According to the authors, reducing the uncertainty in strength assessment may be achievable by utilizing Fuzzy signatures.

Fuzzy signatures (FSig.) have been used successfully for describing civil engineering problems with uncertain input parameters (see e.g. [2, 3]). The aim of this research was to determine the strength of a masonry wall, which is defined by the compressive strength of masonry structural material. An adequate fuzzy aggregation (fitting the area's international standards) was created with measured and estimated parameters as arguments, where the compressive strength of masonry structural material is the root—vertex v_0—of the fuzzy signature. The grouping between the signature's leaves (arguments) was created with experts based on the connections between the arguments. A hierarchically structured structure was generated as a rooted tree forming the signature, by representing the subgroups as sub-trees. The first step of the examination is always the visual inspection without instrumented measurements—it will be the first sub-tree of the FSig, one input of the weighted geometric mean aggregation. This will give a first overview of the state of the masonry, but its limitations are that the results strongly depend on the subjectivity of the observer. To describe the findings, only linguistic expressions are used, which then were transformed into fuzzy sets and were so used in order to create one part of the fuzzy signature.

The next step will be the application of non-destructive measurements, such as Schmidt-hammer, sonic tomography, and penetration tests—all producing sub-trees of the root aggregation.

These tests are used to determine the parameters with specific numbers. They give an opportunity for additional measurements to make the results of the evaluation more reliable and more accurate. The operations between the measured (uncertain) values interpreted as fuzzy numbers—although there was not a specific formula which could describe the impact of leaves to their subtree's root—were done with multiplication based on the extension principle [4] and n-th root of fuzzy numbers [5] as weighted geometric mean in most of the cases. "*A main disadvantage of the multiplicative operation in this case is that by multiplication, the shape of L-R type fuzzy numbers (so triangular or trapezoidal numbers)* [6] *is not preserved. In many situations this problem is solved by approximating the result of the extension principle-based multiplication by a triangular or trapezoidal number.* [7]" Although it has a disadvantage, we used the multiplication for the root operation, because it has well elaborated methods for fuzzy numbers.

The aim of this research was to create a scale based on uncertainties of measured results and the observer's impressions which describes the compressive strength of masonry properly with the use of non-destructive testing methods.

In Sect. 2, we discuss the application of fuzzy arithmetic in the aggregations in the FSigs (cf. [1]) on the fuzzy numbers modeling the compressive strength measurements and observations on the masonry structures. In Sect. 3, we overview the results and the future plans of this research.

2 Applications of Fuzzy Numbers in Determination of the Compressive Strength of Masonry

Paper [8] gives guidelines concerning methods for estimating the strength of masonry based on instrumental measurements. These instrumental tests may provide valuable information about the state of masonry walls and are suitable for determining compressive strength of its components, namely, bricks and mortar. The resulting strength based on the measured values of brick's, mortar's and masonry wall's condition is defined as a weighted geometric mean

$$k * \sqrt[10]{f_b^7 * f_m^3} \quad (1)$$

where k is a coefficient—input parameter with a maximum value of 1—considering the type of brick (solid or hollow core) and purpose of mortar (general or thin layer) based on Sonic Tomography test, f_b is the brick's strength value obtained from the Schmidt hammer test, and f_m is mortar's strength value calculated from the Penetration test.

The measured parameters and values based on subjective expert opinion were transformed into fuzzy numbers. The most compelling reason for using fuzzy numbers in these calculations was that there all factors are subject to noise, and uncertainty. E.g. by the visual inspection we can only guess that the wall's salt damage is low, medium or high—where low salt damage is the best and high the worst—no exact numbers can be assigned to these values in the relevant expressions. For representing the inexact, fuzzy values of the above linguistic expressions, "most typical" values were determined over the interval [0.5; 1]. These values were assigned to the expressions based on our goals that better conditions get higher results and worse conditions get lower results. Because of these values are just guesses, we did not assign well-defined numbers to the factors, but rather we used trapezoidal fuzzy numbers as approximation like in Table 1 and in Fig. 1.

The overlap sections can be interpreted both as the masonry wall has medium salt damage and that it is highly damaged. It cannot be defined with standard measured values, but it is just an evaluation guess based on an expert's subjective opinion. This gave a reason to use these intermediate, "not unambiguously defined" intervals.

Table 1 Corners of trapezoidal membership functions of salt damage

Salt damage	Membership functions
High	$\left(0.7\ 0.7\ 0.775\ 0.8\right)$
Medium	$\left(0.775\ 0.8\ 0.9\ 0.925\right)$
Low	$\left(0.9\ 0.925\ 1\ 1\right)$

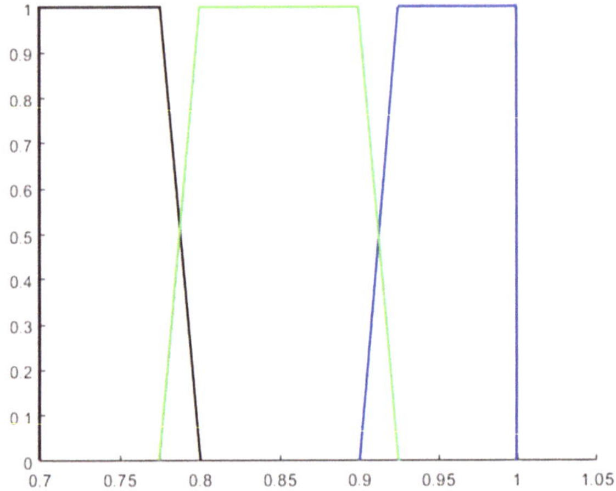

Fig. 1 Salt damage membership function

Based on [9] we created FSig G to model the interrelations among the factors that were used for the estimation of the strength. Signature G may be seen in Fig. 2. Its depth is $d(G) = 4$.

After the determination of the factors, we used weighted geometric mean according to (1) which was given to calculate the compressive strength of the masonry wall based on brick's strength and moisture. We tested our method for a good state wall—low moisture, low salt damage, whitewash—and for a bad condition wall—high moisture, high salt damage, whitewash. We expected that the result will be significantly different for these types which means that we could determine quite exactly the state of masonry wall without destructive testing. For the particular results we got the following membership functions which belong to Visual inspection (Fig. 3) and Instrumental test (Fig. 4). The results of sub-tree's roots are to be found in Table 2.

The final results for the Masonry strength represented as fuzzy numbers are $MS_{high} = \left(1.7467\ 1.8484\ 2.8652\ 3.0085\right)$ and $MS_{low} = \left(1.0689\ 1.1241\ 1.7558\ 1.8590\right)$, which are represented in Fig. 5. It can be seen that there is some overlap between a high quality and a low quality masonry, but the cores have no intersections. By increasing the number of initial settings, probably there would be some overlap which strengthens the guess that we cannot determine exactly, whether a wall is strong enough, but we can get at least a good approximation as the result.

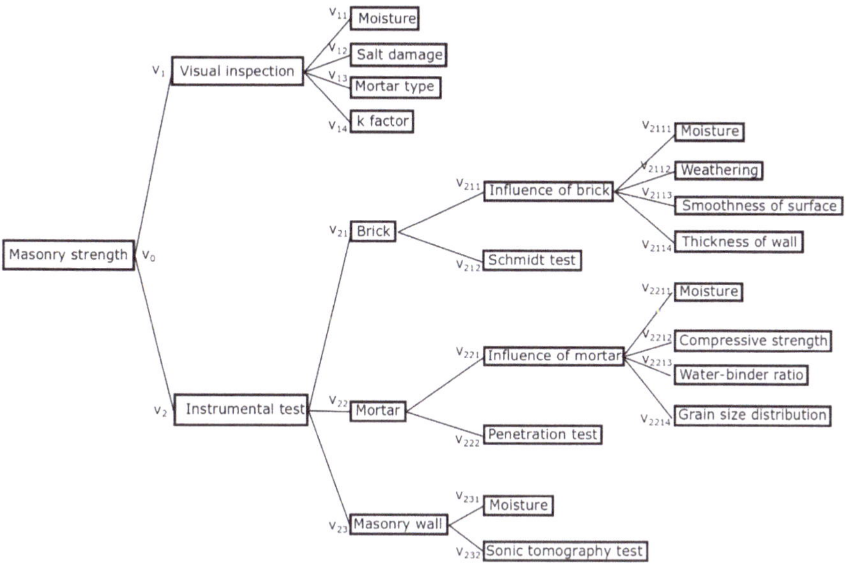

Fig. 2 Fuzzy signature of the inspections and the measurements

Fig. 3 Visual inspection's membership f

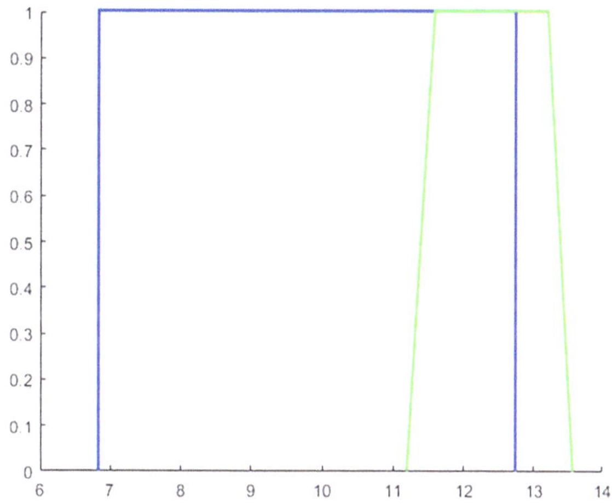

3 Conclusion and Future Work

In line with preliminary expectations, by varying the input parameters, different values were obtained to characterise the strength of the masonry. This provides useful additional information for the qualitative assessment of masonry in practice. But if more tests could be done, there could be more overlap even in the cores of the fuzzy numbers representing similar masonry walls. Thus, it would be more difficult

Fig. 4 Instrumental test's membership f

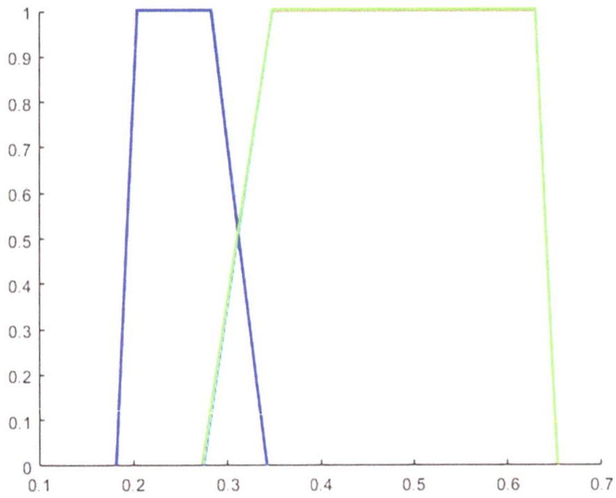

Table 2 Different states of subtree roots described with the corners of trapezoidal membership functions

Membership function of	Good condition wall	Bad condition wall
Brick	$\left(11.2008\ 11.5904\ 13.2077\ 13.572\right)$	$\left(6.8399\ 6.8399\ 12.7366\ 12.7366\right)$
Mortar	$\left(0.2751\ 0.3487\ 0.6300\ 0.6538\right)$	$\left(0.1822\ 0.2045\ 0.2840\ 0.3420\right)$
Masonry wall	$\left(0.9196\ 0.9533\ 1\ 1\right)$	$\left(0\ 0\ 0.7080\ 0.7439\right)$
Visual inspection	$\left(0.7793\ 0.8711\ 1\ 1.0283\right)$	$\left(0.5422\ 0.5913\ 0.6999\ 0.7440\right)$
Instrumental test	$\left(3.6532\ 4.0320\ 5.3012\ 5.4638\right)$	$\left(2.2860\ 2.3752\ 4.0694\ 4.3029\right)$

Fig. 5 Masonry strength's membership function

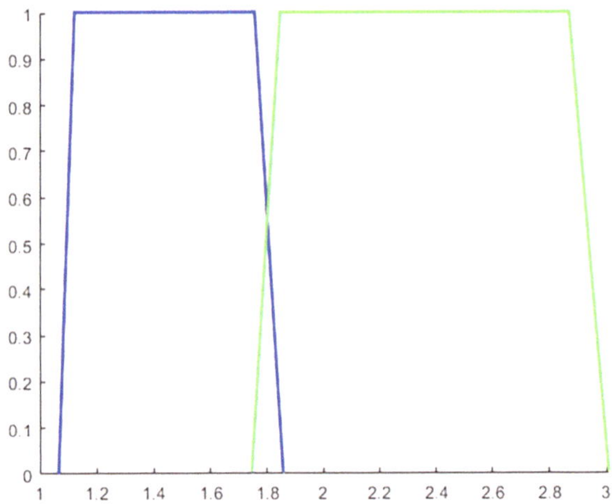

to make a difference between good and bad condition masonry walls. On the other side, we could determine the missing factors and their respective weights, thus the results would be more accurate.

On the other hand, with more test results we could determine more accurate values for unique properties which can make narrower fuzzy intervals and less overlaps. Because of the many subjective parameters, definitely more results are required to determine the exact weights correctly in the operations in the model. The results of such extended measurements may lead to efficient tuning of the parameters by machine learning to enhance the method.

In this study, we used only approximations for the operations with fuzzy numbers, so we always applied trapezoidal numbers, which are easy to interpret and to investigate, but of course, other types of fuzzy numbers could be used, however, these would be less easy to interpret. It was shown by numerous experiments in the literature that there is only a slight difference between the trapezoidal approximation and of more sophistically shaped membership function models' defuzzified results, so, the extension of membership function shapes would not influence the validity significantly, while calculations and interpretability are much easier this way.

In the results, the interpretation of the overlapping sections is obviously uncertain, in these regions, there is a need for more tests. It seems a good solution for this problem to introduce more parameters and then set up a new, more complex, "cascaded" signature to make the results clear.

References

1. Kóczy, L.T., Cornejo, M.E., Medina, J.: Algebraic structure of fuzzy signatures. Fuzzy Sets Syst. **418**, 25–50 (2021)
2. Molnárka, G., Kóczy, L.T.: Decision support system for evaluating existing apartment buildings based on fuzzy signatures. Int. J. Comput. Commun. Control **6**(3), 442–457 (2011)
3. Bukovics, Á.: Structural health analysis and fuzzy signature based intervention decision support modelling. Ph.D. thesis, Győr (2012)
4. Zadeh, L.A.: Fuzzy sets. Inform and Control **8**, 338–353 (1965)
5. Elmoasry, A., Shams, M., Yaqoob, N., Kausar, N., Gaba, Y.U., Rafiq, N.: Numerical scheme for finding roots of interval-valued fuzzy nonlinear equation with application in optimization. J. Funct. Spaces (2021)
6. Goguen, J.: L-fuzzy sets. J. Math. Anal. Appl. **18**, 145–174 (1967)
7. Bede, B., Fodor, J.: Product type operations between fuzzy numbers and their applications in geology. Acta Polytechnica Hungarica, 123–139 (2006)
8. EN 1996-1-1:2005, Eurocode 6: Design of masonry structures—Part 1-1: General rules for reinforced and unreinforced masonry structures. European Committee for Standardization, Brussels
9. Dormány, A., Orbán, Z.: Experimental investigation of the sonic velocity in historical masonry walls. Pollack Periodica Int. J. Eng. Inf. Sci., 12 (2023). Print ISSN 1788-1994. Online ISSN 1788-3911

MIX
Papier aus verantwortungsvollen Quellen
Paper from responsible sources
FSC® C105338

If you have any concerns about our products,
you can contact us on
ProductSafety@springernature.com

In case Publisher is established outside the EU,
the EU authorized representative is:
Springer Nature Customer Service Center GmbH
Europaplatz 3, 69115 Heidelberg, Germany

Printed by Libri Plureos GmbH
in Hamburg, Germany